Introduction to Dynamics

James W. Dally

William L. Fourney

University of Maryland at College Park

College House Enterprises, LLC
Knoxville, Tennessee

The manuscript was prepared with 11 point New Times Roman font using the PC version of Microsoft Word. The Word.doc files were converted to pdf files using Adobe Acrobat Pro DC. The printed version of the book is "printed on demand" by Ingram-Spark. The CD version of the book was prepared in a protected pdf file with a password for opening it.

About the front cover photo

The front cover shows a frame from a high speed video record of a sled test conducted at the Naval Air Weapons Station at China Lake, CA. The rocket propelled sled, shown at the lower part of the photo has moved from a level track to a ramp that curves downward near the end of the track. The sparks from the sled feet are generated by the large forces required to change the orientation of the sled.

The projectile is a BLU 116 bomb used to penetrate a series of concrete targets located down range. A bomb fuze is located in the rear end of the projectile. The test was conducted to monitor the performance of the fuze as the BLU 116 engaged a series of concrete targets located down range from the end of the tracks.

A close inspection of the photo shows the collars that held the BLU 116 to the rocket sled flying into the air. An even closer inspection shows the bolts that fastened the collars to the sled flying into the air. These were the explosive bolts that were detonated just as the sled began to move down the ramp near the end of the track.

College House Enterprises, LLC.
email   jwd@collegehousebooks.com
Visit our web site at http://www.collegehousebooks.com

13 Digit ISBN 978-1-935673-47-7

Bill and Jim both dedicate this book to:

education in general and to higher education in particular,

which provide a pathway to a better and more fulfilling life.

ABOUT THE AUTHORS

James W. Dally (Jim) obtained a Bachelor of Science and a Master of Science degree, both in Mechanical Engineering from the Carnegie Institute of Technology. He obtained a Doctoral degree in Mechanics from the Illinois Institute of Technology. He has taught at Cornell University, Illinois Institute of Technology, the U. S. Air Force Academy and served as Dean of Engineering at the University of Rhode Island. He is currently a Glenn L. Martin Institute Professor of Engineering (Emeritus) at the University of Maryland, College Park.

Dr. Dally has also held positions at the Mesta Machine Co., IIT Research Institute and IBM. He is a Fellow of the American Society for Mechanical Engineers, Society for Experimental Mechanics, and the American Academy of Mechanics. He was appointed as an honorary member of the Society for Experimental Mechanics in 1983 and elected to the National Academy of Engineering in 1984. Professor Dally was selected by his peers to receive the Senior Faculty Outstanding Teaching Award in the College of Engineering and the Distinguish Scholar Teacher Award from the University. He was also a member of the University of Maryland team receiving the 1996 Outstanding Educator Award sponsored by the Boeing Co. He received the Daniel C. Drucker Medal from the ASME in 2012 and the Archie Higdon Distinguished Educator Award from the Mechanics Division of the ASEE in 2013.

Professor Dally has co-authored several other books: *Experimental Stress Analysis, Experimental Solid Mechanics, Photoelastic Coatings, Instrumentation for Engineering Measurements, Packaging of Electronic Systems, Mechanical Design of Electronic Systems, Statics, Mechanics of Materials, Production Engineering and Manufacturing, and Introduction to Engineering Design, Books 1, 2, 3, 4, 5, 6, 7, 8, 9, 10, 11 and 12.* He has authored or coauthored about 200 scientific papers and holds five patents.

William L. Fourney (Bill) obtained his BS degree in Aerospace Engineering from West Virginia University, his MS in Theoretical and Applied Mechanics from West Virginia University, and his PhD in Theoretical and Applied Mechanics from University of Illinois – CU. He has taught courses at all levels within the Mechanical Engineering Department at the University of Maryland and also entry level courses in the Aerospace Engineering Department and within the A. James Clark School of Engineering's Engineering Science Program at the University of Maryland. He has held the administrative positions of Chairman of Mechanical Engineering and Chairman of Aerospace Engineering. He was most recently Associate Dean of the A. James Clark School of Engineering for Undergraduate Studies from which he resigned in September of 2018. He was also the lead Professor of the Keystone Program until September of 2018. The Keystone Program is unique at the University of Maryland and places only professors in the program that want to teach freshmen and sophomore engineering classes in such a way to assist in every way possible the success of students taking those classes. He conducts research in the area of dynamic experimental mechanics, specializing in explosive loading. He has authored or coauthored approximately 280 scientific papers, reports, and book chapters. He was awarded the Max M. Frocht Award and the Charles E. Taylor Award by the Society for Experimental Mechanics. He holds 2 patents.

DEDICATION

ABOUT THE AUTHORS

PREFACE

CONTENTS

LIST OF SYMBOLS

PART I KINEMATICS OF PARTICLES AND RIGID BODIES

CHAPTER 1   BASIC CONCEPTS IN MECHANICS

CHAPTER 2   RECTILINEAR MOTION OF A PARTICLE

CHAPTER 3   MOTION OF A PARTICLE IN THREE DIMENSIONS

CHAPTER 4   KINEMATICS OF RIGID BODIES IN PLANAR MOTION

PART II KINETICS OF PARTICLES AND RIGID BODIES

CHAPTER 8 WORK AND ENERGY: RIGID BODIES

CHAPTER 9 MOMENTUM AND IMPULSE: PARTICLES AND RIGID BODIES

PREFACE

This book an "Introduction to Dynamics" is intended to be used by sophomore or junior students after they have taken Statics and possibly Mechanics of Materials. Comparing the content of Dynamics to that of Statics shows that Dynamics is a much more difficult course. Dynamics requires students to use mathematical tools that include:

        1. Vector cross and dot products
        2. Vector algebra
        3. Law of sines and law of cosines
        4. Chain and product rules of differentiation
        5. Differentiation and integration of elementary functions
        6. Area integrals of simple functions
        7. Calculus and algebra in Cartesian, tangential and cylindrical coordinates

In addition Dynamics includes many different concepts involving energy, momentum and impulse, including conservation of all three of these quantities. The student must learn to solve for linear and angular velocities and accelerations in three different coordinate systems. All in all Dynamics is a challenging course for students to master and a challenging course for instructors to effectively teach.

The challenges offered by a dynamics course afford the student an excellent opportunity to significantly expand their knowledge and to learn how to address difficult and significant problems.

Dynamics is divided into two different topics; kinematics and kinetics. Kinematics only involves the geometry of the path of an object, which may be a particle or a two or three dimensional rigid body. Forces are not involved in kinematic analyzes. Kinematics deals with the motion of bodies without regard to the forces that are required to produce that motion. The student will use differential and integral calculus in solving kinematics problems.

Kinetics is more complex, because the student must consider the forces acting on the particle or rigid body and use Newton's second law to determine its acceleration, velocity and position. In addition concepts such as energy, momentum, impulse and conservation of these quantities are employed to solve certain types of problems.

Over the years the length and page size of almost all textbooks has expanded significantly. For example one of the authors studied both Statics and Dynamics from the same book titled "Analytical Mechanics for Engineers, Third Edition" by Fred B. Seely and Newton E. Ensign. The book was printed in black and white with a page size of 5.5 in. by 8.5 in and had a page count of 450 pages. The figures were all line drawings and they were numbered consecutively. No figures based on photographs are found in the text. There were a few illustrated problems and many homework problems a few of them were provided with answers.

Today we find the more traditional Mechanics textbooks (for either Statics or Dynamics), available from publishers such as Pearson, McGraw-Hill or Prentice Hall, have 600 to 700 or more pages printed with many shades of color. We believe the traditional textbooks are too expensive, too involved, and too long and tend to overwhelm some students. We decided to write an Introductory Dynamics book with many fewer pages and a more direct approach. We reduced page count in a number of ways.

1. We eliminated the homework problems that account for hundreds of pages in the traditional Dynamics books. We have learned that solutions to any problem in any book is available online for a modest cost. The ready availability of solutions eliminates the value of the homework problems in any textbook. Instead of creating 25 to 30 problems for each major topic, we have increase to number of examples in each chapter to illustrate problem solution methods for the students.
2. We have eliminated chapters that are rarely covered in a first course in Dynamics such as free or forced vibrations, and greatly reduced coverage of three dimensional kinematics and kinetics of rigid bodies.
3. We have eliminated the large number of photographs that are found in traditional Dynamics books. These photos enhance the presentation, but do nothing to explain how to solve dynamics problems for the students.
4. We have eliminated the answer section, because we previously eliminate the homework problems.
5. The traditional Dynamics textbooks cover kinematics and kinetics of particle motion first and then repeat the coverage for rigid bodies later in the book. This requires repeating some information, because of the time between the two presentations. In this Dynamics textbook, we alternate between particles and rigid bodies in adjacent chapters. This presentation eliminates repetition of the content.

This book differs in that chapters 2 through 9 open with a concept problem to encourage the student's thought processes before introducing them to a multitude of equations. We believe the students with background in Physics courses and early Mechanics courses will be able to follow these basic concept problems prior to the introduction of the basic equations required for their solution. The concept problems are followed by non-technical discussion dealing with the topic upon which the concept problem is based. The technical discussions deal with current events that are related to the technical content in the chapter and are intended to couple the course content with real life.

The book is divided into two parts:

PART I KINEMATICS OF PARTICLES AND RIGID BODIES

The book has an introductory chapter and three Chapters that cover the kinematics of particle and rigid body motion. A brief description of each of the chapters follows.

Chapter 1 gives brief descriptions of the four branches of Mechanics, including Statics, Mechanics of Materials, Dynamics and Fluid Mechanics. A very brief review of the history of mechanics is given that shows that the laws governing solid mechanics have been well understood for many years. Newton's three laws of motion that form the foundation for mechanics have been described. Newton's laws provide the basis for the equations of motion which dominate the content in Part II. The types of forces are described. Finally, a coverage of basic units and quantities is given.

Chapter 2 introduces the concept of **Kinematics** and indicates that the motion of a particle or rigid body is independent of the forces involved, although the forces could be very large or small. Forces become involved when Kinetic motion is treated in Part II. This chapter deals with the rectilinear motion of a particle, which is the most straight forward topic in the book. However, the topic affords us the opportunity to introduce basic quantities of position, displacement, velocity and acceleration. After acceleration is introduced, three cases of acceleration are considered — a constant acceleration, an acceleration that varies with time and an acceleration that is a function of x. Relations for the equation of motion of a particle are developed for these three cases. Finally numerical methods are introduced where a numerical method for determining time derivatives is presented. Also a method for integrating a function of x to determine the area under a curve is presented. Examples were presented illustrating the method of numerical differentiation and numerical integration. A total of 16 Examples were presented to demonstrate the concepts involved in the study of rectilinear motion of a particle.

Chapter 3 introduces motion of a particle in three dimension. While the title indicates that three dimensional motion is the topic described, most of the development pertains to planar motion, which involves less mathematics. A curvilinear path is prescribed by introducing a position vector in rectangular coordinates. The position vector is differentiated to develop three dimensional equations for velocity and acceleration. Projectile motion is introduced early to provide interesting example of planar motion. Equations for range, and height of the projectile are developed as function of angle of the gun tube, and exit velocity. Two additional coordinate systems are introduced. The normal and tangential system and the radial and transverse system. Both systems are used to determine the accelerations associated with different types of particle motion along curvilinear paths. Angular acceleration is introduced for a particle moving with a velocity v along a curvilinear path. Finally methods were introduced to determine tangential velocity and normal and tangential acceleration for the important case of a particle in motion about a circular path. A total of 17 Examples were presented to demonstrate the concepts involved in the study the motion of a particle along curvilinear paths.

Chapter 4 introduces planar motion of a rigid body. This is the most difficult of the chapters dealing with kinematics, because rigid body is permitted to rotate. Particles on the other hand do not have size; hence, rotation (of a point) is impossible. To address rigid body rotation two coordinate systems are establish — one fixed in space and the second that may translate and rotate. Radial position parameters r_P and r_Q are established in this coordinate system, which are differentiated to yield relations for velocity, acceleration and relative velocity and relative acceleration. This more general theory enables the analysis of the translation of rigid bodies and the rotation of rigid bodies about fixed point. Finally, after the important concept of instantaneous center was introduced, its effectiveness in the solution of several Examples was demonstrated. A total of 11 Examples were presented to demonstrate the concepts involved in the study the planar motion of a rigid bodies.

PART II KINETICS OF PARTICLES AND RIGID BODIES

The book has five Chapters that cover the kinetics of particle and rigid body motion. A brief description of each of the chapters follows.

Chapter 5 introduces the motion of a particle when subject to a force or forces. Newton's second law provides the basic equation that couples the applied system of forces to the acceleration. The equations of motion for planar rigid body motion are references to three different coordinate systems — rectangular, normal and tangential and radial and transverse. Equations for the velocity, position and displacement are derived for these coordinates. A total of nine Examples were presented to demonstrate the concepts involved in the study the planar motion of a particle.

Chapter 6 introduces the motion of a rigid body when subject to a system of forces. Again Newton's law provides the equation of motion. However Newton's law contains the mass moment of inertia I for rotation. Methods were introduced to develop relations for determining the mass moment of inertia in both two and three dimensional bodies. A table was provided summarizing results for the mass moment of inertia for the shapes of rigid bodies commonly found in applications. The parallel axis theorem was introduced and used to determine the mass moment of inertia in composite bodies. After developing methods for determining the mass moments of inertia, methods for determining the velocity and acceleration of rigid bodies in translation, rotation and combined translation and rotation were developed. A total of 16 Examples were presented to demonstrate the concepts involved in the study the planar motion of rigid bodies.

Chapter 7 deals with the concepts of work and energy associated with particle motion. The development follows the usual approach of introducing easily understood definitions and equations and moving on to the more difficult concepts. Accordingly the definition and equation of work and/or energy associated with the application of a constant force to a particle in rectilinear motion is introduced. Then work and/or energy due to a particle undergoing curvilinear motion is presented. Kinetic energy and then potential energy are defined. The concept of changing from one form (kinetic) energy to another form (potential) energy and vice versa is introduced. Finally power and efficiency of the work performed on particles is discussed. A total of 12 Examples were presented to demonstrate the concepts involved in the study of work and energy associated with particle motion.

Chapter 8 covers the concepts of work and energy associated with rigid body motion. Kinetic energy for a rigid body is treated by considering the body translating and rotating that gives rise to equations involving both linear velocity v and angular velocity ω. The coverage of work and/or energy due to applied forces is similar to that introduced in the previous chapter; however the coverage of work and/or energy due to the application of moments to a rigid body while new is not difficult. The most important concept introduced is that of conservation of energy. It is used in many examples in this and the subsequent chapters. A total of seven Examples were presented to demonstrate the concepts involved in the study of work and energy associated with rigid body motion.

Chapter 9 deals with the concepts of momentum and impulse for both particles and rigid bodies. Previous discussion of Newton's second law covered linear momentum. This chapter introduces methods for dealing with angular momentum. Linear impulse is introduced and it is related to the change in the linear momentum. Angular momentum is determined from mass moment of inertia and the angular velocity of a rigid body. The concept of angular impulse and its relationship to the change in angular momentum is introduced. Finally, the concept of impact is introduced that enables the analysis of impacting bodies. Because impact is not elastic, energy is lost and the conservation principles cannot be employed. The approach is to introduce the coefficient of restitution (an experimentally determined quantity) to develop a method of analysis for determining velocities of impacting bodies. Analysis methods for solving problems involving direct central impact and eccentric impact are covered. A total of 12 Examples were presented to demonstrate the concepts involving momentum and impulse associated with both particle and rigid body motion.

ACKNOWLDEGEMENT

Dr. Patrick McAvoy carefully reviewed the entire text and made many suggestions for improvements in presentation and for corrections. We thank him very much for his expertise and the many hours that he devoted to improving this book.

James W. Dally
William L. Fourney
University of Maryland at College Park
Spring, 2019

SYMBOLS

$\mathbf{a}$ linear acceleration (vector)
a linear acceleration (scalar)
a_n normal acceleration
a_r radial acceleration
a_t tangential acceleration
a_T transverse acceleration
A area

d/dt durative operator
d distance

e coefficient of restitution
$\mathbf{\mathcal{E}}$ efficiency
E energy
E_k kinetic energy
E_P potential energy
E_W work energy

ft feet
$\mathbf{F}$ force (vector)
F force (scalar)

G center of gravity, universal
      gravitational constant
g gravitational constant
g_e gravitational constant

$\mathbf{H}$ angular momentum (vector)
h height
$\mathbf{HP}$ horsepower

IC instantaneous center
$\mathbf{I}$ linear impulse (vector)
$\mathbf{IA}$ angular impulse (vector)
I mass moment of inertia
$\mathbf{i}$ unit vector

J joule
$\mathbf{j}$ unit vector

kg kilogram
k spring rate

k_T torsional spring rate
$\mathbf{k}$ unit vector

L length

m mass
m meter
$m\mathbf{v}$ linear momentum (vector)
mv linear momentum (scalar)
M mass of the Earth, moment
$\mathbf{M}$ linear momentum (vector)

N newton
n normal direction, coordinate

P power
p pressure

Q point identifier

$\mathbf{r}$ radius (vector)
r radius (scalar)
r radial direction, coordinate
R radius (scalar)
R_e radius of the Earth
$\mathbf{R}$ reaction force (vector)
R reaction force (scalar)

s second, displacement,
     distance
SI International System of
      Units

t tangential direction,
     coordinate
t time
T transverse direction,
     coordinate

$\mathbf{u_a}$ unit vector
$\mathbf{u_\beta}$ unit vector
$\mathbf{u_r}$ unit vector
$\mathbf{u_T}$ unit vector

$\mathbf{u_v}$ unit vector

v_t tangential velocity
$\mathbf{v}$ velocity (vector)
v velocity (scalar)
V volume

$\mathbf{W}$ watt
W weight

x coordinate, direction

y coordinate, direction

z coordinate, direction

α angular acceleration
β angle
Δ difference
γ angle
ρ radius
θ angle
ω angular velocity

CHAPTER 1

BASIC CONCEPTS IN MECHANICS

1.1 INTRODUCTION

The subject of mechanics is usually divided into four different courses, which include:

1. Statics

Statics, your first course in mechanics, involved forces acting on bodies in equilibrium. You learned that forces were vectors, which required knowing their magnitude as well as their directions. You learned how to draw free body diagrams and used them to aid you in writing the equations of equilibrium. You used the fact that the body was at rest or traveling at a constant velocity in a straight line to determine the forces required to keep the body in equilibrium.

2. Dynamics

As an engineer you need to learn how to determine the forces acting on bodies that are not in equilibrium. These bodies may be speeding up or slowing down and they may not be traveling in a straight line (so changing direction). Dynamics is divided into two different types of analysis; kinematics and kinetics. Kinematics only involves the geometry of the path of an object, which may be a particle or a two or three dimensional body. Forces are not involved in kinematic analyzes. Kinematics deals with the motion of bodies without regard to the forces that are required to produce that motion. You will use differential and integral calculus in solving these kinematics problems.

Kinetics is more complex because you must consider the forces acting on the body and use Newton's second law to determine acceleration, velocity and position of particle or a rigid body. In addition concepts such as energy balances, momentum, and impulse are employed to solve certain types of problems.

3. Mechanics of Materials

When you study mechanics of materials, the deformation of the body is an essential consideration in the analysis. The external forces are determined by static or dynamic analyzes. In Mechanics of Materials, you assume that the deformations of the body are small and plane sections remain plane during the deformation process. This assumption enables you to determine the distribution of internal forces in the body. You use these internal forces with simple equations ($\sigma = P/A$ and $\sigma = My/I$) to determine the stresses at critical locations in the body. The material from which the body is fabricated is of critical importance in the solution of many problems in Mechanics of Materials for two reasons. First, the deformations of the body due to the forces are markedly affected by the rigidity of the material (its elastic modulus). Second, whether the body fails or not depends on its strength, which is a physical property of the material. Finally, you compare the results of your analysis for stresses with the strength of the material and calculate the margin of safety for the structure.

4. Fluid Mechanics

In fluid mechanics the situation is entirely different, because the body under consideration is not rigid. The body is either a gas or a liquid or a mixture of the two fluids. The deformations are sufficiently large to be considered as flows. The flow can be compressible (gases under higher pressures) or incompressible (liquids or gases at low pressures). The flow can occur in closed channels or open channels. The flow may be internal to a conduit or external to some surface. The flow may be stable (laminar) or unstable (turbulent). The phase of the material may change during the process and the resulting flow may consist of two-phase mixtures (some liquid and some gas). Because of the complexities inherent in fluid mechanics, the subject usually is studied after a student has established a thorough understanding of the other three branches of mechanics.

Statics and dynamics both deal with rigid bodies that are subjected to a system of external forces. In the traditional study of statics, we are concerned with determining either internal and/or external forces acting on a structural element that is in a state of equilibrium (usually at rest). In dynamics, the forces acting on the body produce motion and the body accelerates or decelerates and may change direction. The analysis in dynamics deals with determining position (displacement), velocity (angular or linear) and acceleration of the body as some function of time. Newton's laws guide our study of mechanics[1]. Consider Newton's second law:

$$\sum \mathbf{F} = \frac{d}{dt}(mv) \qquad (1.1)$$

where $\Sigma \mathbf{F}$ is the sum of all of the forces acting on the rigid body; m is the mass of the body; $\mathbf{v}$ is the linear velocity and d/dt is the derivative operator.

In dividing the study of mechanics into its four subjects, scholars have considered the special situation where the velocity is constant (often zero) and developed the subject we call **statics** based on this simplification of Newton's second law. In this special situation:

$$\sum \mathbf{F} = 0 \qquad (1.2)$$

We study motion in dynamics — the velocity of the rigid body is changing with time ($\mathbf{v} \neq 0$) and the general form of Eq. (1.1) applies. However, in most situations the mass of the rigid body is constant (dm/dt = 0) and if this is the case, Eq. (1.1) reduces to:

$$\sum \mathbf{F} = m\frac{d\mathbf{v}}{dt} = m\mathbf{a} \qquad (1.3)$$

where $\mathbf{a}$ = dv/dt is the linear acceleration of the rigid body.

_____

[1] Sir Isaac Newton (1642-1727) formulated three laws of motion and the law of universal gravitational attraction.

The study of dynamics is usually divided into two broad topics — kinematics and kinetics. Kinematics involves the geometry of the path along which the body is moving. The body may be a particle or it may be a two- or three- dimension rigid body. Forces are not involved in kinematic analyses. Kinematics treats only the geometric relationships between the properties that define a body's motion (position, velocity and acceleration). In kinematics you must know the path along which the body is moving and determine from that motion the accelerations that are required to cause the body to change velocity as well as the accelerations required to change directions of the particle or the rigid body.

Kinetics, on the other hand, involves the forces acting on the particle or rigid body and the resulting position velocity and acceleration. In the study of both statics and dynamics, the material from which the body is fabricated is of no concern provided the body remains essentially rigid under the action of the imposed forces.

Dynamics is a challenging course, because it requires that you use a number of advanced mathematical concepts including:

1. Vector cross and dot products
2. Vector algebra
3. Law of sines and law of cosines
4. Chain and product rules of differentiation
5. Differentiation and integration of elementary functions in one dimension
6. Area integrals of simple functions
7. Calculus and algebra in Cartesian and cylindrical coordinates

The challenges offered by a dynamics course afford you an excellent opportunity to significantly expand your knowledge and to learn how to address difficult and significant problems.

1.2 HISTORY OF MECHANICS

Leonardo da Vinci (1452 – 1519) was a man with incredible talent and one of the great Renaissance masters. He was a painter, sculptor, musician, architect and engineer. At the age of about 30, he served as the principal engineer for the Duke of Milan supervising the construction of bridges and war machines for the Duke's military ventures [1]. Leonardo is not known for his mathematical discoveries, but rather for his engineering innovations that implied his thorough understanding of mechanics.

Sir Isaac Newton (1642 – 1727) was a giant in both mathematics and mechanics. He was only 24 years old when he developed the mathematics known today as differential calculus. Later Newton turned his attention to planetary motion and formulated his famous laws of motion. We will cover these three laws in more detail in the next section. However, these laws are the foundation for much of the content contained in mechanics courses on statics and dynamics. He is best known for his universal law of gravity [2] that explains the attractive forces between two bodies. The gravitational force is an internal force, because it acts on each elemental volume of mass within a body. The attractive forces **F** between two masses, illustrated in Fig. 1.1, are given by:

$$F = G\, m_a\, m_b\, /r^2 \qquad (1.4)$$

where F is the magnitude of the attractive gravitational force; $G = 6.673 \times 10^{-11}$ m^3/(kg - s^2) is the universal gravitational constant; m_a and m_b are the masses of bodies A and B and r is the distance between the center of masses of the two bodies.

Because Newton's law of gravity is extremely important, we will describe it in more detail later in this chapter.

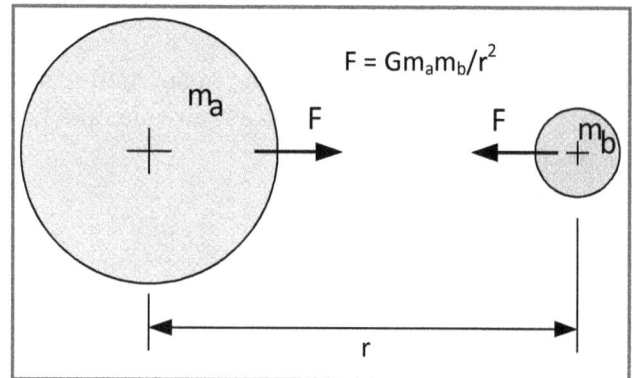

Fig. 1.1 Attractive forces on both bodies are developed by the gravitation field.

The history of mechanics is rich with accomplishments of many mathematicians. As the subject evolved, engineers designing safe efficient structures and machines such as airplanes, automobiles, bridges, skyscrapers, and space satellites reduced the mathematical formulations to practice.

1.3 NEWTON'S LAWS OF MOTION

Sir Isaac Newton wrote three laws of motion that form the foundation for both statics and dynamics. They are:

(1) If the sum of forces acting on a body is zero (i. e., $\Sigma\, \mathbf{F} = 0$), the body will either:

    a.   Remain at rest.
    b.   Move at a constant velocity.

(2) If the sum of the forces $\mathbf{F}$ acting on a body is not zero, the body will undergo a time rate of change of the linear momentum (mv) given by:

$$\sum \mathbf{F} = \frac{d}{dt}(m\mathbf{v}) \qquad (1.1)$$

If the mass m of the body remains constant with respect to time dm/dt = 0, Eq. (1.1) reduces to:

$$\sum \mathbf{F} = m\frac{d\mathbf{v}}{dt} = m\mathbf{a} \qquad (1.3)$$

(3) The force exerted by body A on body B is equal in magnitude but opposite in direction to the force that body B exerts on body A. This third law is sometimes called the law of action and reaction.

1.3.1 Newton's First Law

The first of Newton's laws is written as:

$$\Sigma \mathbf{F} = 0 \tag{1.2}$$

Another way of representing Eq. (1.2) is:

$$\mathbf{F}_1 + \mathbf{F}_2 + \ldots\ldots + \mathbf{F}_n = 0 \tag{1.2a}$$

where n forces are acting on the body.

Equation (1.2), in one form or the other, is used extensively in both statics and mechanics of materials. It is used every time that the **equilibrium** equations are written. To illustrate the meaning of the mathematical symbol $\Sigma\mathbf{F}$, let's examine the drawing of the potato like body presented in Fig. 1.2. We have four forces acting on this body. They are pointed in four different directions and have four different magnitudes. The forces are **vector** quantities that must be characterized by specifying both a magnitude and a direction[2]. When the **vector sum** of these forces is zero, the body is in equilibrium.

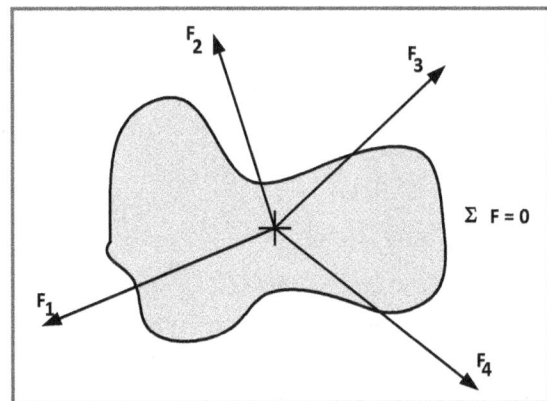

Fig. 1.2 Four forces acting on a body that is in equilibrium if $\Sigma\mathbf{F} = 0$.

1.3.2 Newton's Second and Third Laws

Newton's second law $\Sigma\mathbf{F} = d/dt(m\mathbf{v}) = m\mathbf{a}$ is the equation used most frequently in the study of dynamics. When the summation of forces is not zero, the body of mass m is subjected to an acceleration **a**, which is a vector quantity.

The third law is often called the law of action and reaction. We illustrate the concept of active and reactive forces in Fig. 1.3. In this illustration, a spherical shaped mass m with a weight W rests on the floor at a contact point. The sphere is in equilibrium under the action of two forces. The first force is the weight W due to gravity that acts downward. The second is the reaction force developed at the

---

[2] We use bold font to represent forces as vector quantities when both magnitude and direction are to be specified. Sometimes we consider only the magnitude of the force, and in this case, normal fonts are used to represent the scalar magnitude.

contact point. The reaction force R is equal in magnitude to the weight W, but opposite in direction, because the sphere is at rest (i. e., in equilibrium).

When any two bodies are in contact (i.e. the sphere and the floor), two forces develop at the contact point. These forces are equal in magnitude, opposite in direction and collinear.

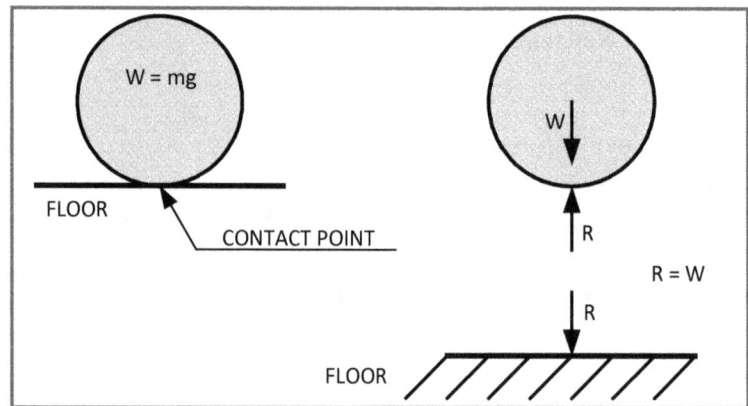

Fig. 1.3 The active force W due to gravity produces a reaction force R at the contact point.

1.4 FORCES

Mechanics involves a study of forces that act on and within a body. In dynamics we consider the external forces acting on a body. Some forces occurring under static (steady state) conditions include:

- Gravitational
- Pressure acting over a defined area
- Elastic (spring)
- Reactions
- Friction
- Magnetic
- Electrostatic

First, let's examine the forces due to gravity, because they are by far the most important. We continuously work and expend huge amounts of energy to overcome gravitational forces. Gravitational forces are the primary concern when we design structures and bridges against failure by collapse or rupture. Even in vehicle design, where other dynamic forces are significant, gravitational forces are critical in the design of both the structure and the power train.

Weight is a force produced by the Earth's gravitational pull on the mass of our body, as illustrated in Fig. 1.4. Suppose we examine the force on a body due to gravity by modifying Eq. (1.5), and letting:

$m_a = m_e$ the mass of the Earth; $m_b = m_b$ the mass of our body and $r = R_e$ the radius of the Earth.

Then we rewrite Eq. (1.4) to give:

$$F = G \, m_e \, m_b \, / R_e^{\,2} \qquad\qquad (1.5)$$

In setting $r = R_e$, we assumed that Earth bound bodies, either those of people or objects, are very small compared to the radius of the Earth, which is 3,960 mi. or 6.37×10^6 m.

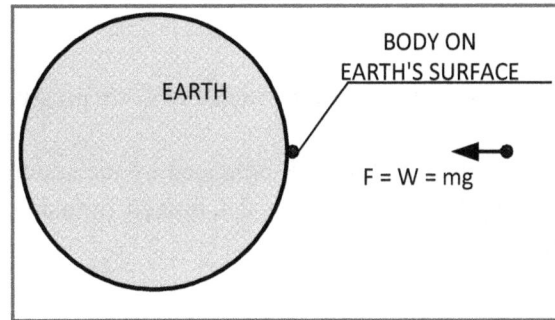

Fig. 1.4 Bodies on Earth's surface are small relative to the Earth's radius.

Next, we collect together all of the quantities in Eq. (1.5) that are constants on Earth, and set them equal to g_e.

$$g_e = Gm_e /R_e^2 \qquad (1.6)$$

Note that g_e is the **gravitational constant** equal to 32.17 ft/s^2 or 9.807 m/s^2. We will drop the subscript in subsequent discussion of the gravitational constant with the understanding that g is to be applied to Earth bound bodies.

Strictly speaking g is not constant, because it varies by a very small amount as we move from one location to another. The Earth is not a perfect sphere, and R_e does not remain constant when we move from Pikes' Peak to Death Valley. However, the variations are so small that we neglect them without introducing significant error in our design analyses.

Combining Eqs. (1.5) and (1.6) gives:

$$F = m_b\, g = W \qquad (1.7)$$

The force F in Eq. (1.7) is the weight W of a body of mass m_b on the Earth's surface.

From our definition, it clear that the units of g are ft/s^2 or m/s^2; hence, g is an acceleration. Indeed, if we jump off of a ladder, our body falls with a constant acceleration of a = g until we hit the ground. Clearly, this relationship is consistent with Newton's second law.

If we travel from Washington, D. C. to Denver, CO, we observe that our weight remains essentially the same. So we get confused and begin to think that the constant in Eq. (1.7) is our weight. It is a reasonable thought, but erroneous. When measuring weight, it is essentially constant if we remain Earth bound. However, the constant quantity in Eq. (1.7) is not the weight W, but the mass m_b. To prove this statement, go to the moon, and measure your weight. It is known that we weigh much less on the moon — about one sixth as much as here on Earth. Since our mass m_b is constant, we weigh less because the gravitational constant for the moon is only about g/6, or $g_m = Gm_m /R_m^2$, based on its mass m_m and radius R_m^2. The smaller gravitational constant for the moon is due to its much smaller mass when compared to the mass of the Earth. The mass of our body is the same whether we are on the moon, Mars the Earth or anywhere in space.

Did you drive your car to the university today? If so, the pressure developed by the combustion of gasoline within the cylinders of your car's engine provided the force to propel it along the roads. Pressure, p acts over an area, A of some surface to create a force that acts normal to that surface.

$$F = p\,A \qquad\qquad\qquad (1.8)$$

where A is the area over which a uniform pressure acts (e.g., the cross sectional area of the piston).

We illustrate the force F produced by the action of the pressure p on the piston, shown in Fig 1.5. The magnitude of the force is determined by using Eq. (1.8); its direction is normal to the surface of the piston.

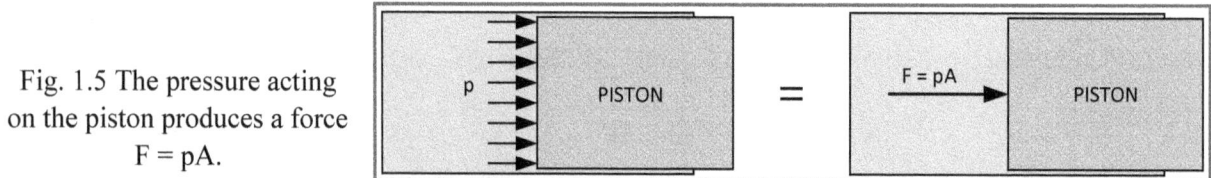

Fig. 1.5 The pressure acting on the piston produces a force $F = pA$.

Another example of the piston and cylinder arrangement is presented in Fig. 1.6, which shows the components of an internal combustion engine including the cylinder and piston. Your automobile has either 4 or 6 of these pistons that power your car as you drive to the university.

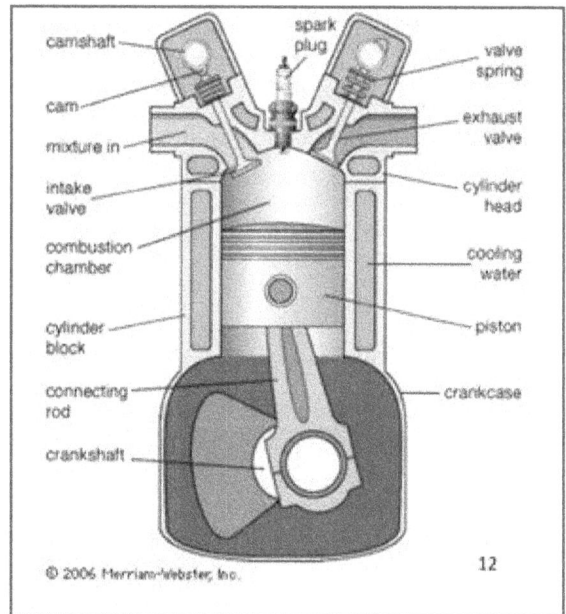

Fig. 1.6 A section cut showing the components in the engine of an automobile. Note the connecting rods and the crankshaft. Why is the cylinder water cooled? Courtesy of Merriman-Webster, Inc.

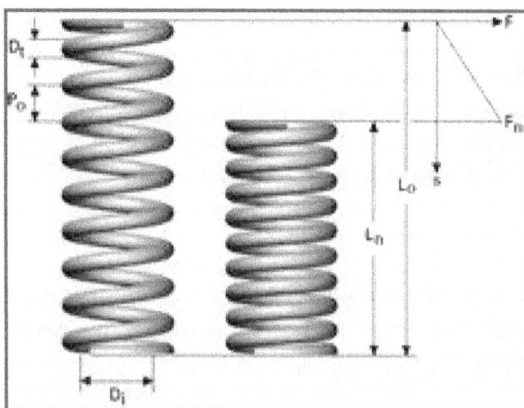

Fig. 1.7 A helical compression spring that is compressed to produce an axial force F linearly dependent on ΔL. Courtesy of Lesjofers, Springs and Pressings.

Springs are also used to provide forces when they are in contact with a body. The force developed is given by:

$$F = k(L_O - L_n) \tag{1.9}$$

where L_O and L_n are defined in Fig. 1.7 and k is the spring rate.

1.5 BASIC QUANTITIES AND UNITS

In the study of mechanics, we will encounter three basic quantities: length, time and mass. These quantities are shown with their respective units for the SI and the U.S. Customary systems of units in Table 1.1. In addition there are many related quantities that we often employ, such as force and acceleration.

Table 1.1
Basic Quantities and Units

System of Units	Length	Time	Mass	Force	g
International System of Units (SI)	meter (m)	second (s)	kilogram (kg)	newton (N), $(kg\text{-}m)/s^2$	9.807 m/s^2
U. S. Customary (FPS)	foot (ft)	second (s)	slug $(lb\text{-}s^2)/ft$	Pound force (lb, $slug\text{-}ft/s^2$)	32.17 ft/s^2

The basic units are not independent because Eq. (1.3) requires that the units be **dimensionally homogenous**. To maintain the dimension homogeneity of Eq. (1.3), we define the units for length, time and mass in the **SI** system, and then derive the remaining basic unit for force, the **newton**, in terms of those units. For the **U. S. Customary** system, the basic unit for mass, the **slug**, is derived in terms of the units for length, time, and force.

In the International System of Units (SI), the length is given in meters (m), the time in seconds (s), and the mass in kilograms (kg). The unit for force is called a newton (N) in honor of Sir Isaac. The newton is derived from Eq. (1.3) so that a force of 1 N will impart an acceleration of 1 m/s^2 to a mass of 1 kg [i. e. 1 N = (1 kg) (1 m/s^2)]. For dimension homogeneity, it is clear that N is equivalent to $(kg\text{-}m)/s^2$. In the SI system the Earth's gravitation constant g = 9.807 m/s^2. With this value of the acceleration due to gravity on Earth, the weight of a mass of 1 kg is:

$$W = mg = (1kg)(9.807 \ m/s^2) = 9.807 \ N$$

In the U. S. Customary System, the length is given in feet (ft), the force in pounds (lb), and the time in seconds (s). The unit for mass is called a slug, which is derived from Eq. (1.3) so that a force of 1 lb will impart an acceleration of 1 ft/s^2 to a mass of 1 slug [i. e. 1 lb = (1 slug)(1 ft/s^2)]. For dimension homogeneity, it is clear that a slug is equivalent to $(lb\text{-}s^2)/ft$. In the U. S. Customary System the gravitation constant g = 32.17 ft/s^2. With this value of the acceleration due to gravity on Earth, the mass of a body weighing 32.17 lb is:

$$m = W/g = 32.17 \ lb/(32.17 \ ft/s^2) = 1 \ slug$$

1.6 SUMMARY

We begin with brief descriptions of the four branches of Mechanics, including Statics, Mechanics of Materials, Dynamics and Fluid Mechanics. A very brief review of the history of mechanics is given that shows that the laws governing solid mechanics have been well understood for many years. Newton's three laws of motion that form the foundation for mechanics have been described. Newton's laws lead to Eqs. (1.2) and (1.3), which are so important that they are repeated here:

$$\Sigma \mathbf{F} = 0 \tag{1.2}$$

$$\Sigma \mathbf{F} = \mathbf{ma} \tag{1.3}$$

Forces are described in considerable detail. Because forces due to gravity and pressure are so common, we provided the equations used to determine their magnitudes. The force or weight due to gravity is:

$$F = m_b\, g = W \tag{1.7}$$

The force due to pressure is:

$$F = p\, A \tag{1.8}$$

REFERENCES

1   Uccelli, A., Leonardo da Vinci, Reynal and Co., New York, NY 1956.
2   Newton, Sir Isaac, Philosophiae Naturalis Principia Mathematica, 1687.

PROBLEMS

We have not included a listing of problems that usually follow the theoretical content of a chapter. After observing the availability of solutions on the internet for essentially all of the homework problems provided in the well-known Mechanics textbooks[3], we have decided not to provide homework problems following each chapter. Instead have increased the number of examples found in each chapter of this textbook. Also we have noticed that many instructors write their own problems patterned after the Examples found in each of the subsequent chapters.

---

[3] See for example, https://www.chegg.com/homework-help.

CHAPTER 2

RECTILINEAR MOTION OF A PARTICLE

2.1   KINEMATICS

To study Kinematics we need to expand our understanding and learn about moving bodies and the accelerations required to produce that movement.   Kinematics only involves the geometry of the path of an object, which may be a particle or a two or three dimensional body. Forces are not involved in kinematic analysis. In kinematics you study how objects (both particles and rigid bodies) move and what that motion implies with regard to the accelerations required to produce that motion.

Kinematics deals with the motion of bodies without regard to the forces that are required to produce that motion. In this chapter we will only deal with particles, which are small entities that may have mass, but not physical dimensions[1]. A particle is infinitesimally small when compared to distance it may travel. Kinematics involves establishing the relations for the displacement, velocity and acceleration of a particle in motion, using time and the geometry of the particle's path.

Later, when we consider the motion of rigid bodies, with meaningful dimensions, the analysis becomes more complex and we must consider the forces acting on the body, energy balances, momentum, impulse, etc. We will treat this class of problems in our study of Kinetics of Rigid Bodies in Part II.

Before beginning the coverage of rectilinear motion, let's consider a concept example that you probably can solve with your current understanding of dynamics from your physics and mathematics courses.

CONCEPT EXAMPLE

Suppose a bullet train operates on a straight level track for a distance of 300 miles to connect Station A and Station B. The train is at rest in Station A, when the operator begins to increase the velocity v of the train at a constant rate of $\Delta v/\Delta t = 1.2$ ft/s^2 until a velocity of 200 MPH is achieved. At this speed, the operator throttles back and holds the train at a constant velocity of 200 MPH until the train is near Station B. As the train approaches Station B, the operator applies regenerative braking and decreases the velocity of the train at a constant rate of $\Delta v/\Delta t = - 0.9$ ft/s^2 so that the train comes to a complete stop at Station B.

We are to determine the following information that characterizes the train's performance.

1.   The time required for the train to increase its velocity from zero to 200 MPH.

---

[1] The bodies modeled by particles may be very large. For example, the Earth is modeled as a particle, when developing the equations for planetary motion.

2. The distance travelled during this time period as the train is accelerating.
3. The time before arriving at Station B, when the operator applies the braking system.
4. The distance travelled during the time period with the regenerative brakes applied..
5. The distance travelled when the train was travelling at the constant speed of 200 MPH.
6. The time the train travelled at the constant speed of 200 MPH.
7. The total time required for the train to travel from Station A to Station B.

Let's begin by constructing a graph that shows the velocity of the train as a function of time. From the problem statement it is clear that the velocity v increases as a constant rate $\Delta v/\Delta t = 1.2$ ft/s^2 from t = 0 to t = t_1. The train then travels at a constant velocity of 200 MPH from t_1 until t_2. At t_2 the brakes are applied and the rate of change of the velocity $\Delta v/\Delta t = -0.9$ ft/s^2. Note this quantity is negative because the velocity is decreasing. The graph is shown below:

The velocity — time profile for the bullet train travelling from Station A to Station B.

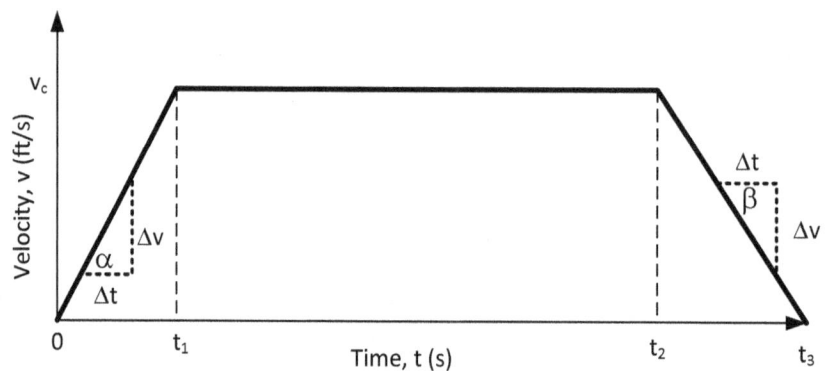

Now that we understand the changes in the velocity with time, we can begin to address the time required for the train to increase its velocity from zero to 200 MPH.

The problem involves units of feet, second and miles per hour (MPH). Let's decide to perform our calculations in feet and seconds. The conversion of 200 MPH into ft/s is shown below:

$$\frac{200\text{mi}}{\text{h}} \times \frac{\text{h}}{3,600\text{s}} \times \frac{5,280\text{ft}}{\text{mi}} = 293.3 \text{ ft / s} \qquad (a)$$

The rate of increase of the velocity with time $\Delta v/\Delta t = 1.2$ ft/s^2, which is an acceleration. We can determine the time t_1 from:

$$t_1 = \frac{v_c}{\Delta v}\Delta t = 293.3 / 1.2 = 244.4 \text{ s} = 4.074 \text{ minutes} \qquad (b)$$

The distance travelled by the train during this period under the constant acceleration of 1.2 m/s^2 is equal to the area under the velocity-time curve from $0 \geq t \leq t_1$, as shown below:

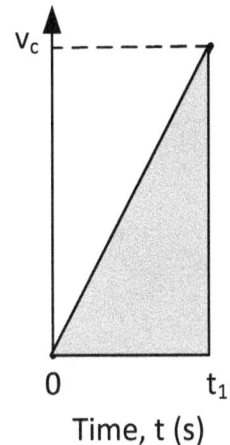

The velocity — time profile for the bullet train as it accelerates from rest to its cruising velocity of 200 MPH.

0   t_1

Time, t (s)

The area of this triangle is:

$$A = ½ \, v_c \, t_1 = s = ½ \, (293.3 \text{ ft/s})(244.4 \text{ s}) = 35{,}841 \text{ ft or } 6.788 \text{ mi.} \qquad (c)$$

Later in the theoretical developments in this chapter, we will derive the equation for the distance s travelled by a particle subjected to a constant acceleration a with an initial velocity v_0 as:

$$s = s_0 + v_0 \, t + ½ \, a \, t^2 \qquad (d)$$

when $s_0 = 0$ and $v_0 = 0$, Eq. (d) reduces to:

$$s = ½ \, a \, t^2 = ½ \, (1.2)(244.4)^2 = 35{,}841 \text{ ft}$$

and $\qquad\qquad s = ½ \, a \, t^2 = ½ \, v_c \, t_1 \qquad\qquad$ which is the area of the triangle shown above.

To determine the time before arriving at Station B, when the operator the operator applies the regenerative braking system, we follow a similar procedure, but note that the velocity is decreasing and the acceleration $\Delta v / \Delta t$ is negative. The rate of decrease in the velocity with time $\Delta v / \Delta t = -0.9 \text{ ft/s}^2$, which is a constant deceleration. We can determine the period of braking from $t_3 - t_2$ as:

$$t_3 - t_2 = \frac{\Delta v}{\Delta v / \Delta t} = -293.3 \, / \, (-0.9) = 325.9 \text{ s} = 5.431 \text{ minutes} \qquad (e)$$

The distance travelled during this time period is given by the area under the velocity time curve shown below:

The area of this triangle is:

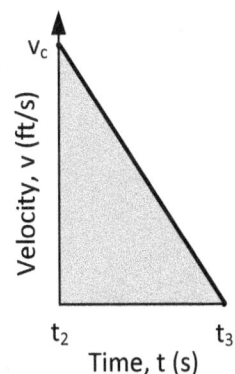

$$A = ½ \, v_c \, (t_3 - t_1) = s = ½ \, (293.3 \text{ ft/s})(325.9 \text{ s}) = 47{,}793 \text{ ft}$$

or 9.052 mi. $\qquad\qquad\qquad (f)$

$t_2 \qquad\qquad t_3$

Time, t (s)

To determine the distance travelled when the train was travelling at the constant speed of 200 MPH, we will use the previous results for distances traveled during the period 0 to t_1 and the period t_2 to t_3. Using these results, we can write:

$$s_2 = s_{Total} - s_1 - s_3 = 300 - 6.788 - 9.052 = 284.2 \text{ mi} \tag{g}$$

The time required to travel 284.2 miles at a constant speed of 200 MPH is:

$$t_2 - t_1 = s/v_c = 284.2/200 = 1.427 \text{ h or } 85.25 \text{ minutes}$$

The total time required for the train to travel from Station A to Station B is:

$$t_{Total} = t_1 + (t_2 - t_1) + (t_3 - t_2) = 4.074 + 85.25 + 5.431 = 94.75 \text{ minutes} \tag{h}$$

Nothing was very complicated in this problem; you noted acceleration multiplied by time yields the increase or decrease in velocity and velocity multiplied by time yields the distance travelled.

DISCUSSION

In the problem statement we have specified an acceleration of 1.2 ft/s and a deceleration of -0.9 ft/s². Are these reasonable accelerations for our bullet train considering human factors.

Under normal conditions, your body must maintain a blood pressure of 22 millimeters of mercury to transport blood from your heart to your brain. Each additional g of upward acceleration (blood flows from the head to the feet) that a person experiences multiplies that requirement. A person's circulation system must maintain 44 mm of blood pressure at 2g, 66 mm at 3g, etc. At accelerations 4 to 5 gs, most people will lose conscious due to oxygen starvation, because their blood stays in their feet and a sufficient quantity cannot reach the brain.

Clearly the accelerations cited in the problem statement are modest compared to those values that result in loss of conscious. What other factors should we consider? Stability is a factor, because people move around on trains to find their seat or go to the rest room. What acceleration will topple a person who is standing in the aisle of an accelerating train? Let's perform a simple analysis to determine the acceleration that will cause someone to lose balance.

We represent the person with a rectangular block, as shown below. The weight of the person is mg and is acting downward from the person's center of gravity. The tipping force is the inertia force equal to ma. If we sum moments about point A due to these two forces, we obtain:

$$\Sigma M_A = mgw/2 - ma_{Max} h/2 = 0 \tag{a}$$

or
$$a_{Max} = gw/h \tag{b}$$

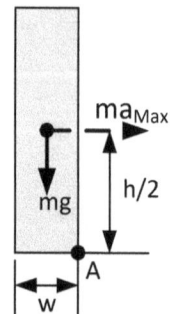

Let's select h = 6 ft, w = 1.5 ft and g = 32 ft/s^2 to represent our person. Then we find:

$$a_{Max} = (32)(1.5/6) = 8 \text{ ft/s}^2$$

Comparing this result with the acceleration and deceleration specified in the problem statement, shows that we have a comfortable margin of safety (8/1.2) = 6.7 for the train to achieve its velocity and for it to brake and come to a controlled stop.

MORE DISCUSSION

While there has been discussion of high speed trains in the U. S. for decades, very little progress has been made. The policy in the U. S. is for passenger trains and freight trains to use the same track. Passenger trains are operated by Amtrak and they operator over tracks owned largely by railroad companies like Norfolk Southern or Northern Pacific. There are several reasons why little progress in developing high speed rail transport has been made in the U. S. and there is no realistic reason to expect progress in the near future.

In Amtrak's 2016 Fiscal Year, the company transported 31.3 million passengers, which averages about 85,700 people per day. With more than 300 Amtrak trains in service, Amtrak services more than 500 stations in 46 states. While these numbers sound like those of an up and coming company building its customer base that is not the case for Amtrak. Since its founding in 1970, the company has never generated a profit. The federal government provides annual subsidy to support passenger service in the U. S. For fiscal 2017 Amtrak requested $1.7 billion, which included $649 million for operating expenses, $920 million for capital construction costs and $263 million in grants.

Federal regulators limit the speed of trains with respect to the signaling method used. Passenger trains are limited to 59 MPH and freight trains to 49 MPH on track without block signal systems. Trains without an automatic cab train stop signal or train control system may not exceed 79 MPH. This regulation has been in effect since 31 December 1951. The regulation was issued the Interstate Commerce Commission following a severe 1946 crash in Naperville, Illinois involving two Chicago, Burlington and Quincy Railroad trains. Following a 1987 train collision in Maryland, freight trains operating in enhanced-speed corridors have been required to have locomotive speed limiters to forcibly slow trains rather than simply alerting the operator with in-cab signals.

STILL MORE DISCUSSION

Mayor Rahm Emanuel announced recently that Chicago has selected Elon Musk's Boring Company to build and operate an "express service to transport people to O'Hare Airport from downtown in 12 minutes on electric vehicles in underground tunnels". The Boring Company indicated that the Chicago Express Loop is to be 100 percent privately funded and the Mayor added that there would be no public subsidy to support it.

Currently, some 20,000 air travelers move between the O'Hare Airport and Chicago's central business district every day, according to the Chicago Infrastructure Trust. The agency adds that the number is predicted to reach to at least 35,000 daily air passengers in 2045.

The Boring Company stated that it would transport passengers on autonomous electric skates traveling at 125-150 miles per hour. Each skate, capable of carrying between 8 and 16 passengers, would be based on the Tesla Model X automobile, and confined to a concrete track. An artist's rendition of a skate is illustrated below.

2.2   RECTILINEAR MOTION OF A PARTICLE

When a particle moves along a straight line, we call this motion rectilinear. Consider the particle, shown in Fig. 2.1, located on the x axis at position P, which is a distance d_i from the origin O.

Fig. 2.1 A particle located on the x axis.

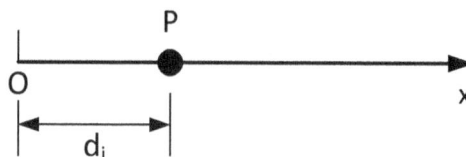

Suppose the particle moves in the positive x direction to a new location Q, as illustrated in Fig. 2.2.

Fig. 2.2 The particle moves from position P to Q along the x axis.

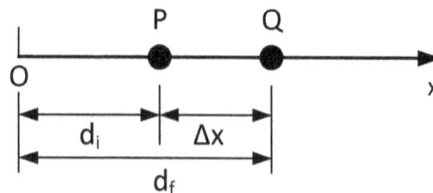

We will assume that the particle was at position P at time t and moves to position Q at $t = t + \Delta t$. We can then write an equation for the instantaneous velocity v as:

$$v = \lim_{\Delta t \to 0} \frac{\Delta x}{\Delta t} \qquad (2.1)$$

Taking the limit for Δt, the relation for the instantaneous velocity v as a time derivative of x is given by:

$$v = \frac{dx}{dt} \qquad (2.2)$$

The average velocity over the time interval Δt is given by:

$$v_{Ave} = \frac{\Delta x}{\Delta t} \qquad (2.3)$$

The velocity v is positive if the position Q has a larger value of x than position P, which makes Δx a positive number (i.e. $d_f \geq d_i$ in Fig. 2.2). On the other hand, if Q has a smaller value of x than P, then the velocity is negative.

The velocity is a **vector** quantity with both magnitude and direction. However, with rectilinear motion, the path of the particle is constrained to the x axis; hence, we may express the vector form of the velocity as:

$$\mathbf{v} = v\,\mathbf{i} \qquad (2.4)$$

where **i** is the unit vector in the x direction.

EXAMPLE 2.1

Consider a particle undergoing rectilinear motion, with its position along the x axis given by:

$$x = 3 + 4t - t^2 \qquad (a)$$

Determine the expression for the velocity as a function of time and calculate the position of the particle and its velocity for the time period from 0 to 5 seconds in steps of one second. Express the position x in meters (m) and the velocity in meters per second (m/s).

Solution:

Use Eq. (2.2) to write the velocity of the particle as:

$$v = \frac{dx}{dt} = 4 - 2t = 2(2 - t) \qquad (b)$$

Use a spreadsheet to determine the position and velocity of the particle for time varying from 0 to 5 s as:

---

t (sec)	x (m)	v (m/s)
0	3	4
1	6	2
2	7	0
3	6	-2
4	3	-4
5	-2	-6

Note for t > 2 s, we find that the velocity is negative. This fact implies that the particle is moving to the left along the x axis. Positive values for the velocity indicate that the particle is moving to the right and negative values indicate that is moving to the left along the x axis.

A graph showing the velocity as a function of time is shown below:

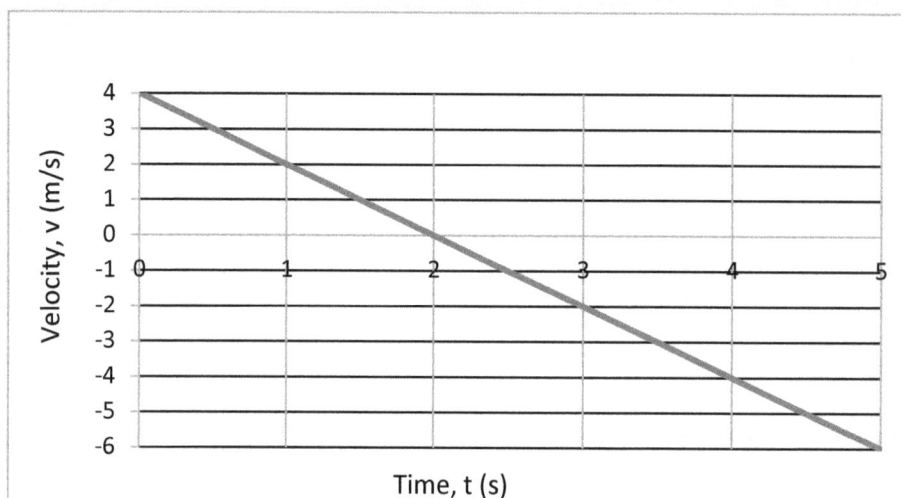

Displacement

As the particle moves along the x axis its position gives its displacement. Consider the results listed above. At t = 0 we found that x = 3 m and at t = 2 s, x = 7 m. Hence, the particle was displaced a distance $d = d_f - d_i = 7 - 3 = 4$ m in this 2 s period of time. We treat the displacement as a positive number, because x increased and $d_f > d_i$ during this 2 s period.

EXAMPLE 2.2

Consider a particle undergoing rectilinear motion, with its position along the x axis given by:

$$x = 2 - 3t + 2t^2 - t^3 \qquad\qquad (a)$$

Determine the equation for the velocity as a function of time and calculate the position of the particle and its velocity for the time period from 0 to 4 seconds. Express the position x in meters (m) and the velocity in (m/s).

Solution:

Use Eq. (2.2) to write the velocity of the particle as:

$$v = \frac{dx}{dt} = -3 + 4t - 3t^2 \qquad\qquad (b)$$

Using a spreadsheet to evaluate this relation yields:

t (sec)	x (m)	v (m/s)
0	2	-3
1	0	-2
2	-4	-7
3	-16	-18
4	-42	-35

A graph showing the velocity as a function of time is shown below:

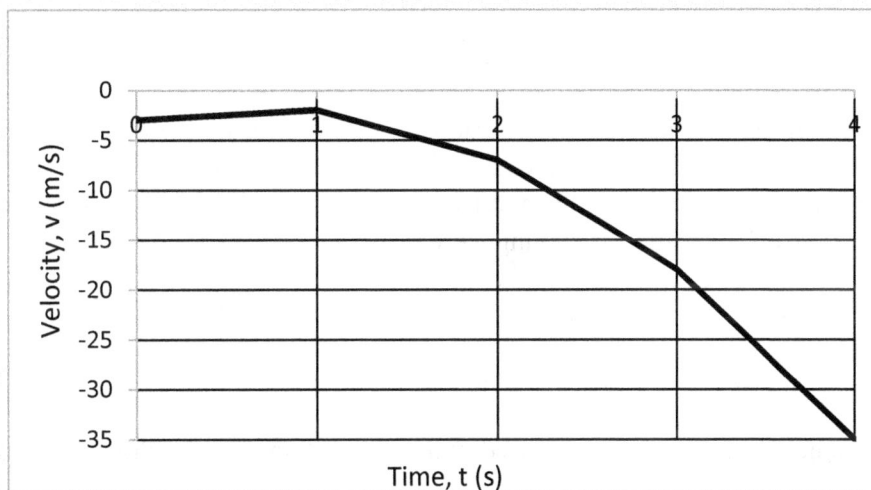

Displacement

These results indicate that the displacement occurring during the two second period (t = 0 to t = 2 s) was minus 6 m, because the position x changed from 2 m to – 4 m. The displacement during this time period is negative, because the particle moved to the left along the x axis.

2.3   ACCELERATION OF A PARTICLE — RECTILINEAR MOTION

Let's consider the particle, shown in Fig. 2.3, which is moving from point P to point Q. As the particle moves from point P to Q, its velocity changes from v at point P to v + Δv at point Q. It is evident that the instantaneous acceleration is given by:

$$a = \lim_{\Delta t \to 0} \frac{\Delta v}{\Delta t} \qquad (2.5)$$

Fig. 2.3 Velocity of the particle changing by Δv during the time increment Δt.

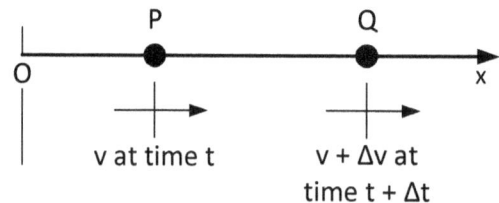

In the limit as Δt goes to zero, the acceleration is given by:

$$a = \frac{dv}{dt} = \frac{d^2x}{dt^2} \qquad (2.6)$$

We can also write the average acceleration as:

$$a_{Ave} = \frac{\Delta v}{\Delta t} \qquad (2.7)$$

For rectilinear motion the direction of the acceleration is known, because the motion is restricted to the x axis. The magnitude of the acceleration is the unknown in this case. The acceleration is positive when Δv > 0 and negative when Δv < 0. When the acceleration is negative, the particle is decelerating (i.e. the velocity is decreasing and the particle is moving more slowly).

The vector notation for the acceleration is:

$$\mathbf{a} = a\,\mathbf{i} \qquad (2.8)$$

Recall Eq. 2.2 and note that it can be written as dt = dx/v. Substitute dt from this result into Eq. (2.6) and simplify it yields:

$$a = v\frac{dv}{dx} \qquad (2.9)$$

EXAMPLE 2.3

Recall EXAMPLE 2.1 with a particle undergoing rectilinear motion, with its position along the x axis given by:

$$x = 3 + 4t - t^2 \qquad \text{(a)}$$

Determine the expression for the acceleration as a function of time and calculate the position of the particle, its velocity and acceleration for the time period from 0 to 5 seconds. Express the position x in meters (m), velocity in meters per second (m/s) and acceleration in meters per second squared (m/s^2).

Solution:

Use Eq. (2.2) to write the velocity of the particle as:

$$v = \frac{dx}{dt} = 4 - 2t \qquad \text{(b)}$$

And use Eq. (2.6) to determine the acceleration as:

$$a = \frac{dv}{dt} = -2 \, m/s^2 \qquad \text{(c)}$$

The results for x, v and a as a function of t from t = 0 to t = 5s, were determined from Eqs. (a), (b) and (c) in a spreadsheet, as shown below:

t (sec)	x (m)	v (m/s)	a (m/s^2)
0	3	4	-2
1	6	2	-2
2	7	0	-2
3	6	-2	-2
4	3	-4	-2
5	-2	-6	-2

EXAMPLE 2.4

Recall EXAMPLE 2.2 that involved a particle undergoing rectilinear motion, with its position along the x axis given by:

$$x = 2 - 3t + 2t^2 - t^3 \qquad \text{(a)}$$

Determine the expression for the acceleration as a function of time and calculate the position of the particle, its velocity and acceleration for the time period from 0 to 4 seconds. Express the position x in meters (m), velocity in (m/s) and acceleration in (m/s^2).

Solution:

Use Eq. (2.2) to write the velocity of the particle as:

$$v = \frac{dx}{dt} = -3 + 4t - 3t^2 \qquad\qquad\text{(b)}$$

Use Eq. (2.6) to determine the acceleration as:

$$a = \frac{dv}{dt} = 4 - 6t \qquad\qquad\text{(c)}$$

The results for x, v and a as a function of t from t = 0 to t = 4s were determined from Eqs. (a), (b) and (c) in a spreadsheet, as shown below:

t (sec)	x (m)	v (m/s)	a (m/s^2)
0	2	-3	4
1	0	-2	-2
2	-4	-7	-8
3	-16	-18	-14
4	-42	-35	-20

In this example, the velocity is negative for the entire time period, and the acceleration is negative for t > 2/3 s.

2.4 DETERMINING THE MOTION OF A PARTICLE MOVING ALONG THE X AXIS

In Section 2.3, we assumed that the equation describing the motion of a particle along the x axis was a known function of time. However, in real applications, we rarely know this function. Instead we often seek to define a suitable function based on our knowledge or measurement of the acceleration of a particle. There are three cases to consider with this alternative approach; the acceleration is (1) a function of time, (2) a function of velocity or (3) a function of position. We consider each of these cases below:

2.4.1 Acceleration Is a Known Function of Time

Let the acceleration $\qquad\qquad a = a(t).$ $\qquad\qquad\qquad\qquad$ (2.10)

Substituting Eq. (2.10) into Eq. (2.6) yields,

$$\frac{dv}{dt} = a(t) \qquad\qquad\text{(a)}$$

or

$$dv = a(t)dt \qquad\qquad\text{(b)}$$

Integrate Eq. (b) with respect to t and obtain:

$$\int_{v_0}^{v} dv = \int_{0}^{t} a(t)dt \qquad (c)$$

where v_0 is the initial velocity of the particle when $t = 0$.

Equation (c) is integrated to obtain:

$$v - v_0 = \int_{0}^{t} a(t)dt \qquad (d)$$

Next reference Eq. (2.2) and write:

$$dx = v\, dt \qquad (e)$$

To obtain a relation for the position x, substitute the results of integrating Eq. (d) into Eq. (e) to obtain.

$$\int_{x_0}^{x} dx = \int_{0}^{t} \left[v_0 + \int_{0}^{t} a(t)dt \right] dt \qquad (2.11)$$

Let's first consider the case of a particle subjected to a constant acceleration with respect to time in EXAMPLE 2.5. This is a very important case, because gravitational acceleration that acts on everything is a constant provided the bodies are Earth bound.

EXAMPLE 2.5

Let's consider a particle subjected to a constant force, which in turn produces a constant acceleration a_0. Determine the equations defining the motion of the particle in terms of the velocity v and the position x as a function of time.

Solution:

Let's begin with Eq. (2.6), as shown below:

$$\frac{dv}{dt} = a_0 \qquad (a)$$

Integrating Eq. (a) yields:

$$\int_{v_0}^{v} dv = a_0 \int_{0}^{t} dt \qquad (b)$$

Completing the integration gives:

$$v = v_0 + a_0\, t \qquad (2.12)$$

---

where v_0 is the initial velocity at $t = 0$, when the acceleration was applied to the particle.

Next we write Eq. (2.2) to determine the position x of the particle as a function of time.

$$\frac{dx}{dt} = v = v_0 + a_0 t \tag{c}$$

Integrating Eq. (c) yields:

$$\int_{x_0}^{x} dx = \int_{0}^{t} (v_0 + a_0 t) dt \tag{d}$$

Completing the integration gives:

$$x = x_0 + v_0 t + a_0 t^2/2 \tag{2.13}$$

where x_0 is the position of the particle at $t = 0$ when the acceleration a_0 was applied.

Another useful relation can be derived by combining the results of Eqs. (2.12) and (2.13). Begin by squaring both sides of Eq. (2.12) and rearranging terms to obtain:

$$\tfrac{1}{2}(v^2 - v_0^2) = a_0 v_0 t + \tfrac{1}{2} a_0^2 t^2 = a_0 (v_0 t + \tfrac{1}{2} a_0 t^2) \tag{e}$$

From Eq. (2.13) note that:

$$\tfrac{1}{2} a_0 t^2 = x - x_0 - v_0 t \tag{f}$$

Substitute Eq. (f) into Eq. (e) and simplify to obtain:

$$v^2 = v_0^2 + 2a_0(x - x_0) \tag{2.14}$$

While the relation given in (2.14) appears to be independent of time, it is not. The position parameter x in Eq. (2.13) depends on time; hence, the time dependence is implicit.

EXAMPLE 2.6

Let's consider a particle that is moving with a constant velocity v_c along the x axis. Determine the position x for this particle as a function of time.

Solution:

Write Eq. (2.2) as:

$$\frac{dx}{dt} = v_c \tag{a}$$

Integrate Eq. (a) to obtain:

$$\int_{x_0}^{x} dx = v_c \int_{0}^{t} dt \tag{b}$$

Completing the integration gives:

$$x = x_0 + v_c t \qquad (2.15)$$

This relation shows that the position x increases as a linear function of the constant velocity v_c.

2.4.2 Acceleration is a Known Function of Position

Let's suppose that the acceleration of the particle is a known function of position x given by a = f(x). Then from Eq. (2.9) we write:

$$v\, dv = a\, dx = a(x)\, dx \qquad (a)$$

Integrate Eq. (a) to obtain:

$$\int_{v_0}^{v} v\, dv = \int_{x_0}^{x} a(x)\, dx \qquad (b)$$

Complete the integration to obtain:

$$\frac{1}{2}\left(v^2 - v_0^2\right) = \int_{x_0}^{x} a(x)\, dx \qquad (2.16)$$

Equation (2.16) enables the determination of the velocity, if the acceleration is a known function of x [i.e. a = f(x)] We demonstrate how to use this relation in Example 2.7.

EXAMPLE 2.7

Consider the acceleration of a particle given by a = a(x) = 3 + 2x and determine the velocity of the particle as a function of x, as x varies from 0 to 4 ft The initial velocity of the particle is v_0 = 0 ft/s.

Solution:

Substituting these parameters into Eq. (2.16) yields: $\frac{1}{2}\left(v^2 - v_0^2\right) = \int_{x_0}^{x} a(x)\, dx$

$$\text{and } v^2 = 0 + 2\int_0^4 (3 + 2x)\, dx \qquad (a)$$

Integrating Eq. (a) gives:

$$v^2 = 2\left[3x + x^2\right]_0^4 \qquad (b)$$

Evaluating Eq. (b) at x=4 ft yields:

$$v^2 = 2[12 + 16] = 56 \text{ ft}^2/\text{s}^2$$

and

$$v = 7.483 \text{ ft/s} \qquad (c)$$

From Eq. (b) it is easy to show that the velocity of the particle increases from 0 to 2.828 ft/s, to 4.472 ft/s to 6 ft/s to 7.483 ft/s as x increases from 0 to 4 ft in steps of 1.0 ft.

2.4.3 Acceleration is a Known Function of the Particle Velocity

Let's suppose that the acceleration of the particle is a known function of its velocity, given by $a = a(v)$. Then from Eq. (2.6) we write:

$$\frac{dv}{dt} = a(v) \qquad \text{or} \qquad dt = \frac{dv}{a(v)} \qquad \text{(a)}$$

From Eq. (2.2) we write:

$$v = \frac{dx}{dt} \qquad \text{or} \qquad dt = \frac{dx}{v} \qquad \text{(b)}$$

Results from Eq. (a) and (b) give:

$$v\frac{dv}{dx} = a(v) \qquad \text{or} \qquad dx = \frac{vdv}{a(v)} \qquad \text{(2.17)}$$

To determine the velocity v with respect to t, integrate Eq. (2.2). To determine the velocity as a function of position x integrate the second relation in Eq. (2.17).

EXAMPLE 2.8

Consider the acceleration of a particle given by $a = a(v) = 1/v$ and determine the position and velocity of the particle as a function of time, as t varies from 0 to 5 s. The initial position of the particle is $x_0 = 0$ and the initial velocity of the particle is $v_0 = 0$ ft/s.

Solution:

From Eq. (2.6) we write:

$$\frac{dv}{dt} = a(v) = \frac{1}{v} \qquad \text{(a)}$$

Integrating Eq. (a) yields:

$$\int_0^t dt = \int_{v_0}^v vdv \qquad \text{(b)}$$

Performing the integration and letting t_0 and v_0 go to zero gives:

$$v = \sqrt{2t} \qquad \text{(c)}$$

Recall Eq. (2.2) and write:

$$dx = v\, dt \qquad \text{(d)}$$

Integrating Eq. (d) yields:

$$\int_{x_0}^x dx = \int_0^t vdt \qquad \text{(e)}$$

Substituting Eq. (c) into Eq. (e) gives:

$$\int_{x_0}^{x} dx = \sqrt{2}\int_{0}^{t} t^{1/2} dt \tag{f}$$

Performing the integration and letting $x_0 = 0$ gives:

$$x = \frac{\sqrt{2}}{3/2} t^{3/2} = \frac{2\sqrt{2}}{3} t^{3/2} \tag{g}$$

Evaluating Eqs. (c) and (g) for x and v with respect to time in a spreadsheet for t increasing from 0 to 5 s in 1.0 s intervals gives the results shown below.

t (sec)	v (ft/s)	x (ft)
0	0.000	0.000
1	1.414	0.943
2	2.000	2.667
3	2.449	4.899
4	2.828	7.542
5	3.162	10.541

Before moving on to the next topic, let's consider a few more examples to demonstrate the methods used to determine the parameters x, v and a, which define the motion of a particle.

EXAMPLE 2.9

Consider a particle undergoing rectilinear motion with its position along the x axis given by:

$$x = 3t - 2t^2 + t^3 \tag{a}$$

Determine the expression for the velocity and acceleration as a function of time and calculate the position of the particle and its velocity for the time period from 0 to 8 seconds. Express the position x in (ft) and the velocity in (ft/s) and the acceleration in (ft/s^2).

Solution:

Use Eq. (2.2) to write the velocity of the particle with respect to time as:

$$v = \frac{dx}{dt} = 3 - 4t + 3t^2 \tag{b}$$

Use Eq. (2.6) to determine the acceleration with respect to time as:

$$a = \frac{dv}{dt} = -4 + 6t \qquad \text{(c)}$$

Substitute numerical values for t into Eqs. (b) and (c) in a spreadsheet to determine the position, velocity and acceleration for time varying from 0 to 8 s:

t (s)	x (ft)	v (ft/s)	a (ft/s^2)
0	0	3	-4
1	2	2	2
2	6	7	8
3	18	18	14
4	44	35	20
5	90	58	26
6	162	87	32
7	266	122	38
8	408	163	44

EXAMPLE 2.10

Let's consider a ball dropped from the roof of a building 15 m high. The ball is subjected to a gravitational force, which results in a constant acceleration downward of 9.81 m/s^2. Determine the velocity v and the position x as a function of time. Determine the time of impact and the velocity of the ball at impact.

Solution:

Let's begin with Eq. (2.6) as shown below:

$$\frac{dv}{dt} = a_0 = 9.81 \text{ m/s} \qquad \text{(a)}$$

Integrating Eq. (a) yields:

$$\int_{v_0}^{v} dv = a_0 \int_{0}^{t} dt \qquad \text{(b)}$$

Completing the integration gives:

$$v = 9.81\, t \qquad \text{(c)}$$

where $v_0 = 0$, because the ball was initially at rest. Note the velocity is increasing as a linear function of time.

Next we write Eq. (2.2) to determine the position x of the particle as a function of time.

$$\frac{dx}{dt} = v = 9.81\,t \qquad\qquad\qquad (d)$$

Integrating Eq. (c) yields:

$$x = 9.81\,(t^2/2) \qquad\qquad\qquad (e)$$

where $x_0 = 0$ is the position of the ball measured from the roof top.

Impact occurs when x = 15 m. Substituting this value into Eq. (e) enables the calculation of the time of impact as:

$$t^2 = \frac{2 \times 15}{9.81} = 3.058\,s^2 \qquad or \qquad t = 1.749\,s \qquad (f)$$

Next use Eq. (c) to determine the velocity at impact:

$$v = 9.81\,t = 9.81 \times 1.749 = 17.16\ ft/s \qquad\qquad (g)$$

EXAMPLE 2.11

The acceleration of a particle is given by $a = 3 - t^2$. At initiation t = 0 and $v_0 = 0$, but $x_0 = 6$ ft. Determine the time when the velocity is again zero. Also determine the position and velocity when t = 2 s. Finally, determine the total distance traveled by the particle as t increases from 0 to 6 s.

Solution:

From Eq. (2.6) we write:

$$a = \frac{dv}{dt} = 3 - t^2 \qquad\qquad\qquad (a)$$

Integrating with respect to t and letting $v_0 = 0$ gives:

$$v = 3t - t^3/3 \qquad\qquad\qquad (b)$$

Set v = 0 in Eq. (b) and solve for t, which is the time for the velocity to decrease to zero.

$$v = 3t - t^3/3 = 0 \qquad or \qquad t^2 = 9 \qquad and \qquad t = \pm 3\,s \qquad (c)$$

At t = 2 s, we note from Eq. (b) that v = 6 – 8/3 = 3.333 ft/s.

Note dx/dt = v; hence, reference Eq. (b) and write:

$$dx/dt = 3t - t^3/3 \qquad\qquad\qquad (d)$$

Integrate Eq. (d) to obtain:

$$x - 6 = 3t^2/2 - t^4/12 \qquad \text{(e)}$$

Using Eq. (b) and (e) we evaluate v and x for t going from 0 to 6 in steps of 1.0 s, as shown below:

t (sec)	v (ft/s)	x (ft)
0	0.000	6.000
1	2.667	7.417
2	3.333	10.667
3	0.000	12.750
4	-9.333	8.667
5	-26.667	-8.583
6	-54.000	-48.000

The total distance travelled by the particle is obtained from the table above as:

$$(12.75 - 6)_{t = 0 \text{ to } 3s} + (12.75 + 48)_{t = 3s \text{ to } 6s} = 67.5 \text{ ft.} \qquad \text{(f)}$$

EXAMPLE 2.12

A particle in rectilinear motion is subject to an acceleration given by $a = 5 - x^2$ with units of feet for position x and ft/s^2 for the acceleration. The particle is initially at rest and located on the origin of the x axis. Determine the velocity v when x = 2 ft, the position when the velocity decreases to zero, and the position of the particle when the velocity is a maximum.

Solution:

From Eq. (2.9) we write:

$$v\frac{dv}{dx} = a = 5 - x^2 \qquad \text{(a)}$$

Integrate Eq. (a) to obtain:

$$\int_{v_0=0}^{v} v\,dv = \int_{x_0=0}^{x} (5 - x^2)\,dx \qquad \text{(b)}$$

Complete the integration to obtain the relation for v as a function of x.

$$v = \left[10x - \frac{2x^3}{3} \right]^{1/2} \qquad \text{(c)}$$

The velocity v when x = 2 ft is given by Eq. (c) as:

$$v = \left[2(10) - \frac{2(2)^3}{3} \right]^{1/2} = 3.830 \, \text{ft/s} \qquad \text{(d)}$$

The position when the velocity goes to zero is calculated by letting v in Eq. (c) go to zero to obtain:

$$v = \left[10x - \frac{2x^3}{3} \right]^{1/2} = \left[\frac{30x - 2x^3}{3} \right]^{1/2} = 0 \qquad \text{(e)}$$

It is evident that:

$$x^2 = 15 \qquad \text{and} \qquad x = 3.873 \, \text{ft} \qquad \text{(f)}$$

and v = 0 at x = 0, because the particle was at rest initially (t = 0).

To determine the position of the particle when the velocity is a maximum, we differentiate Eq. (c) with respect to x, set the result equal to zero and solve for x. Differentiating Eq. (c) yields

$$\frac{dv}{dx} = \frac{1}{2} \left[10x - \frac{2x^3}{3} \right]^{-1/2} \left[10 - 2x^2 \right] = 0 \qquad \text{(g)}$$

Inspection shows that the two values of x that solve Eq. (g) are:

x = $\sqrt{5}$ = 2.236 ft that causes the second bracket term to vanish and gives v = 3.861 ft/s, which is the maximum value for the velocity of the particle.

and

x = $\sqrt{15}$ = 3.873 ft that causes the first bracket term to vanish and gives v = 0, which is the minimum value of the particle's velocity.

2.5 RELATIVE MOTION OF TWO PARTICLES MOVING ALONG THE X AXIS IN RECTILINEAR MOTION

Consider two particles P and Q both moving along the x axis, as shown in Fig 2.4. The positions of the particles P and Q, both measured from the origin O, are given by x_P and x_Q respectively. The distance between the two particles is $x_{Q/P}$, which is given by:

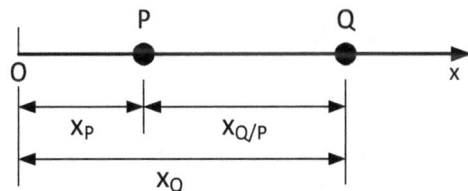

Fig. 2.4 Two particles P and Q both moving along the x axis.

$$x_{Q/P} = x_Q - x_P \qquad \text{(2.18a)}$$

or

$$x_Q = x_{Q/P} + x_P \qquad \text{(2.18b)}$$

It is evident from Fig. 2.4 that $x_{Q/P}$ is positive when $x_Q > x_P$ and negative otherwise.

The relative velocity of the particle Q with respect to the particle P is established by differentiating Eq. (2.18) with respect to time to obtain:

$$v_{Q/P} = v_Q - v_P \qquad (2.19a)$$

or

$$v_Q = v_{Q/P} + v_P \qquad (2.19b)$$

The relative velocity $v_{Q/P}$ is positive when $v_Q > v_P$ otherwise it is negative.

The relative acceleration of the particle Q with respect to the particle P is established by differentiating Eq. (2.19) with respect to time to obtain:

$$a_{Q/P} = a_Q - a_P \qquad (2.20a)$$

or

$$a_Q = a_{Q/P} + a_P \qquad (2.20b)$$

The relative acceleration $a_{Q/P}$ is positive when $a_Q > a_P$ otherwise it is negative.

EXAMPLE 2.13

Car P and car Q are traveling in the same direction on an interstate highway. Car Q is travelling at a constant velocity of 60 MPH (with a = 0) in the right hand lane. Car P is traveling at a velocity of 50 MPH but is accelerating at 3 ft/s². Determine the time for car P to overtake car Q. Also determine the velocity of car P, when it overtakes car Q. Note in Fig. E2.13 that car Q is 100 ft ahead of car P.

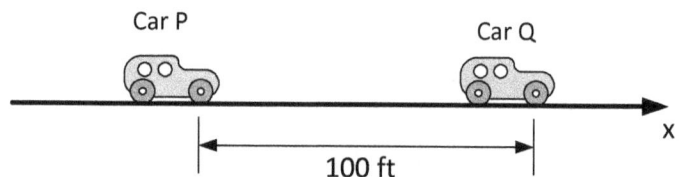

Fig. E2.13 Car P in the process of overtaking car Q.

Solution:

Let's first convert the velocity of the two cars from MPH to ft/s.

$$50 \frac{mile}{hour} \times \frac{hour}{3600s} \times \frac{5,280ft}{mile} = 73.33 \, ft/s \qquad 60 \frac{mile}{hour} \times \frac{hour}{3600s} \times \frac{5,280ft}{mile} = 88.0 \, ft/s$$

Now that we have the velocities of the two cars in suitable units, let's define the position of car P at t = 0 as $x_P = 0$; hence it is clear that $x_Q)_{t=0} = 100$ ft.

We also know from the problem statement that car Q is travelling at a constant velocity of 88 ft/s and is not accelerating. Accordingly we can write the relation for the position of car Q at time t as:

$$x_Q = 88 \, t + x_Q)_{t=0} = 88 \, t + 100 \, ft \qquad (a)$$

The velocity v_P of car P is changing with time, because it is accelerating. The velocity v_P at time t is:

$$v_P = (v_P)_{t=0} + a_P \, t \tag{b}$$

Integrating Eq. (b) with respect to time gives:

$$x_P = (v_P)_{t=0} \, t + \tfrac{1}{2} \, a_P \, t^2 \tag{c}$$

or

$$x_P = 73.33 \, t + 1.5 \, t^2 \tag{d}$$

Car P overtakes car Q when $x_P = x_Q$. Then equating Eqs. (a) and (d) gives:

$$73.33 \, t + 1.5 \, t^2 = 88.0 \, t + 100 \tag{e}$$

Rearranging Eq. (e) in its quadratic form yields:

$$1.5 \, t^2 - 14.67 \, t - 100 = 0 \tag{f}$$

The two roots of Eq. (f) are:

$$t = 14.41 \text{ s and } t = -4.626 \text{ s} \tag{g}$$

The negative root is impossible; hence, the time for car P to overtake car Q is 14.41 s.

The velocity of car P when it overtakes car Q is given by Eq. (b) as:

$$v_P = (v_P)_{t=0} + a_P \, t = 73.33 + 3(14.41) = 116.6 \text{ ft/s} = 79.47 \text{ MPH} \tag{h}$$

EXAMPLE 2.14

The car is travelling at a constant velocity of 50 MPH, when it is passed by a truck traveling at a constant velocity of 60 MPH. When the truck is 90 ft in front of the car, the car driver accelerates at 3 ft/s^2 and passes the truck with a separation distance, shown in Fig. E2.14. Determine the time required to pass the truck with the 80 ft margin shown in Fig. E2.17. Also determine the speed of the car at this time.

Fig. E2.14

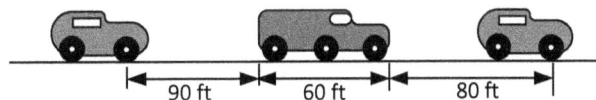

|← 90 ft →| |← 60 ft →| |← 80 ft →|

Solution:

Let's begin by converting the velocities of the car and truck from MPH to ft/s as:

$$\frac{60 \text{mile}}{\text{h}} \times \frac{\text{h}}{3,600\text{s}} \times \frac{5,280\text{ft}}{\text{mile}} = 88\text{ft} / \text{s} \qquad 50 \text{ MPH} = 73.33 \text{ ft/s} \tag{a}$$

The velocity of the car after it begins to accelerate is given by:

$$v_{Car} = v_0)_{Car} + at \qquad (b)$$

The velocity of the truck is a constant:

$$v_{Truck} = v_0)_{Truck} = 88 \text{ ft/s} \qquad (c)$$

The position of the car is:

$$x_{Car} = x_0 + v_{Car})_0\, t + \tfrac{1}{2} at^2 \qquad (d)$$

and

$$x_{Car} = 0 + 73.33\, t + 1.5 t^2 \qquad (e)$$

The position of the truck relative to the car is:

$$x_{Truck} = x_0)_{Truck} + v_{Truck})_0\, t \qquad (f)$$

and

$$x_{Truck} = 90 + 88t \qquad (g)$$

Taking into account the length of the truck and the margin after the passing is complete enables us to write:

$$x_{Car} - x_{Truck} = 60 + 80$$

Substituting from Eqs. (e) and (g) gives:

$$73.33\, t + 1.5t^2 - 90 - 88t = 60 + 80$$

$$1.5\, t^2 - 14.67\, t - 230 = 0 \qquad (h)$$

Solving the quadratic equation for time t and selecting the realist result gives:

$$t = 18.20 \text{ s} \qquad (i)$$

The velocity of the car at this time is:

$$v_{Car} = 73.33 + 3(18.20) = 127.93 \text{ ft/s} \qquad \text{or} \qquad 87.22 \text{ MPH}$$

Let's consider a more incremental approach to generating the solution of EXAMPLE 2.14 by dividing the solution into two parts—the time t_1 for car P to come abreast of the truck Q and then the time t_2 to complete the pass by creating a separation of 80 ft.

Solution:

Begin by defining the position of the car P as $x_P = 0$ and the position of the truck $x_Q = 60 + 60 = 150$ ft at $t_1 = 0$.

The truck Q is travelling at a constant velocity of 88 ft/s with an acceleration $a_Q = 0$. Its position as a function of time t_1 is:

$$x_Q = 150 + v_Q t_1 = 150 + 88t_1 \qquad (a)$$

The position of car P as a function to time t_1 is:

$$x_P = v_P t_1 + \tfrac{1}{2} a_P t_1^2 = 73.33t + \tfrac{1}{2}(3) t_1^2 \qquad \text{(b)}$$

Car P is abreast (the front wheel of the car is in line with the front of the truck Q) when $x_P = x_Q$; hence, Eqs. (a) and (b) yield:

$$150 + 88t_1 = 73.33t + \tfrac{1}{2}(3) t_1^2 \qquad \text{(c)}$$

Reducing Eq. (c) to a standard quadratic form gives:

$$1.5 t_1^2 - 14.67 t_1 - 150 = 0 \qquad \text{(d)}$$

The quadratic equation solver gives the positive root as:

$$t_1 = 16.02 \text{ s} \qquad \text{(e)}$$

At this time

$$x_Q = 150 + 88(16.02) = 1,560 \text{ ft}$$

$$x_P = 73.33(16.02) + 1.5(16.02)^2 = 1,560 \text{ ft}$$

and

$$v_P = 73.33 + 3 t_1 = 73.33 + 3(16.02) = 121.4 \text{ ft/s} \qquad \text{(f)}$$

Let's next determine the time to generate the 80 ft separation. In this second phase of the pass, let's reset the clock at $t_2 = 0$ when the two vehicles are abreast. The we may write that the pass is complete when:

$$x_P = x_Q + 80 \qquad \text{(g)}$$

At $t = t_2$ the velocity $v_Q = 88 t_2$ and then from Eq. (g) we write $x_P = 88t_2 + 80$ \qquad (h)

However from Eq. (f) we can write:

$$v_P = 121.4 + 3 t_2 \qquad \text{(i)}$$

and

$$x_P = 121.4 t + 1.5 t_2^2 \qquad \text{(j)}$$

Equating Eqs. (h) and (i) and simplifying, we can write the quadratic equation as:

$$1.5 t_2^2 + 33.4 t_2 - 80 = 0 \qquad \text{(k)}$$

The quadratic equation solver gives the positive root as:

$$t_2 = 2.181 \text{ s} \qquad \text{(l)}$$

The total time for the trick Q to pass car P is then given by:

$$t + t = 16.02 + 2.18 = 18.20 \text{ s} \qquad \text{(m)}$$

This result is same as that determined previously in EXAMPLE 2.14.

2.6   NUMERICAL METHODS

In most engineering applications, the position or velocity of a particle or a rigid body are not given as explicit functions of time. Instead sensors are employed in tests of prototypes or models to measure parameters such as position, velocity or acceleration. Numerical methods are then used to process the experimental data to obtain the information characterizing the motion. For example, numerical differentiation is used to convert position (displacement) data into velocity or numerical integration is used to convert velocity data into displacement information.

2.6.1 Numerical Differentiation

When we differentiate a function of say $x = f(t)$, we determine dx/dt. With numerical differentiation, we represent dx/dt with $\Delta x/\Delta t$. If numerical data from a displacement sensor is available, this quantity is easy to calculate. The process to determine Δx and Δt is illustrated in Fig. 2.5, where a curve of displacement x is shown as a function of t. A point of interest is selected along the time t axis. Two additional points an equal distance to the left and right of the point of interest are marked t_1 and t_2. Construct vertical lines from these points to intersect the x - t curve. The vertical construction lines intersect the x-t curve and give the values of y_1 and y_2. Next we subtract the numerical values for y and t to obtain Δy and Δt as:

$$\Delta y = y_2 - y_1 \tag{2.21}$$

and

$$\Delta t = t_2 - t_1 \tag{2.22}$$

then

$$\frac{\Delta x}{\Delta t} = \frac{x_2 - x_1}{t_2 - t_1} \tag{2.23}$$

Finally, the time corresponding to the derivative is given by:

$$t = (t_1 + t_2)/2 \tag{2.24}$$

Fig. 2.5 Graphical illustration of the time derivative of $x = f(t)$.

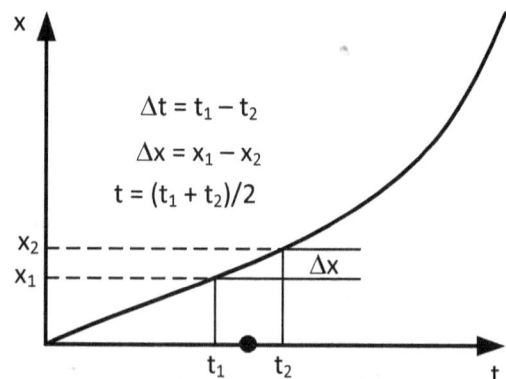

The process can be repeated at any point along the t axis, where the derivative is sought. It is also easy to enter the x-t data into an Excel spreadsheet and to determine the $\Delta x/\Delta t$ over the entire range of data for x as a function of t.

EXAMPLE 2.15

Numerically differentiate the function $x = t^2$ at intervals of 0.1 s over the time range from 0 to 3 s.

Solution:

We have constructed an Excel spreadsheet, with time ranging from zero to 3 s in steps of 0.1 s. We then constructed a second adjacent column giving x according to the relation $x = t^2$. In column C cell C5, we computed Δx from B6 – B4 which corresponds to Δx at x given by B5. Finally, we determined $\Delta x/\Delta t$ in column D cell D5 by dividing according to C5/0.2, because $\Delta x = 0.2$ s. The copy and paste commands in Excel were used to fill the spreadsheet, as shown below:

Derivative Example			
x=t^2		$\Delta t = 0.2$	
t	x	Δx	$\Delta x/\Delta t$
0	0		
0.1	0.010	0.040	0.20
0.2	0.040	0.080	0.40
0.3	0.090	0.120	0.60
0.4	0.160	0.160	0.80
0.5	0.250	0.200	1.00
0.6	0.360	0.240	1.20
0.7	0.490	0.280	1.40
0.8	0.640	0.320	1.60
0.9	0.810	0.360	1.80
1.0	1.000	0.400	2.00
1.1	1.210	0.440	2.20
1.2	1.440	0.480	2.40
1.3	1.690	0.520	2.60
1.4	1.960	0.560	2.80
1.5	2.250	0.600	3.00
1.6	2.560	0.640	3.20
1.7	2.890	0.680	3.40
1.8	3.240	0.720	3.60
1.9	3.610	0.760	3.80
2.0	4.000	0.800	4.00
2.1	4.410	0.840	4.20
2.2	4.840	0.880	4.40
2.3	5.290	0.920	4.60
2.4	5.760	0.960	4.80

2.5	6.250	1.000	5.00
2.6	6.760	1.040	5.20
2.7	7.290	1.080	5.40
2.8	7.840	1.120	5.60
2.9	8.410	1.160	5.80
3.0	9.000		

2.6.2 Numerical Integration

It is easy to perform integration using numerical methods. For example, we may integrate to determine the area under a curve y = f(x) by dividing the area into thin slices, calculating the area of each slice (a rectangle) and then adding all of these areas to obtain the total area. We represent the mathematical operation by a summation sign instead of the more typical integration sign, as shown below:

$$A = \sum_{n=1}^{N} A_n \qquad (2.25)$$

The accuracy of the integration depends on the width of the slices and the height of the rectangle relative to the value of y over the interval Δx. Consider y = f(x) that yields the graph with increasing y with increasing x, as shown in Fig. 2.6. In this illustration, we established the height of the associated with the minimum value of y over the range of the interval Δx. It is evident that this procedure will under estimate the area, as the rectangular areas do not capture the small triangular areas above the rectangles.

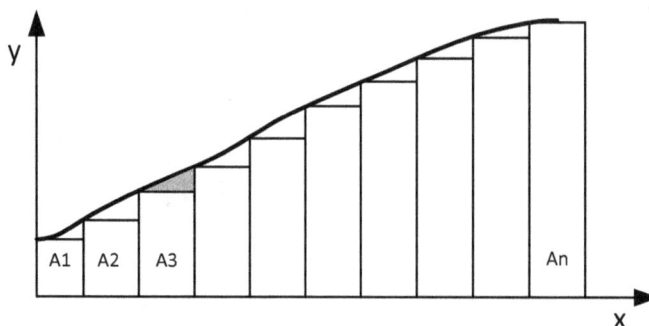

Fig. 2.6 An example of an error in selecting the height of the rectangular areas.

The error demonstrated in Fig. 2.6 is eliminated by adjusting the height of each rectangle to intersect the value for y at the midpoint of the Δx interval, as shown in Fig. 2.7. With this selection for the height of the rectangle, two small triangular areas are formed; one that represents a positive error and the other a negative error. These small areas tend to cancel out and the area estimates using this procedure are sufficiently accurate for most engineering calculations.

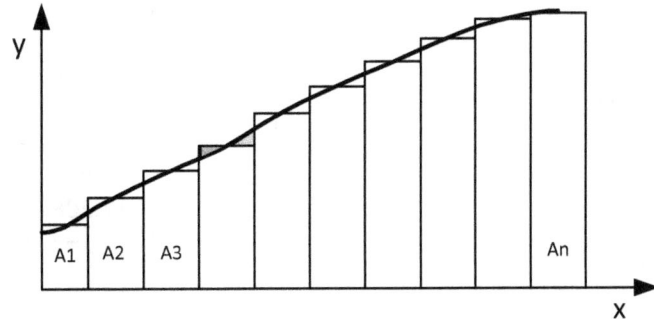

Fig. 2.7 An example of an improved
method for selecting the height of the
rectangular areas.

EXAMPLE 2.16:

Consider the function $v = f(t) = t^3 + 2t^2 + 3t + 4$ and determine the area under its curve over the limits from zero to 2.0 by integrating numerically and calculate the error involved.

$$A = \int_0^2 (t^3 + 2t^2 + 3t + 4)dt$$

Integrate this polynomial to obtain:

$$A = \left[\frac{t^4}{4} + 2\frac{t^3}{3} + 3\frac{t^2}{2} + 4t \right]_0^2 = [16/4 + 16/3 + 12/2 + 4(2)] = 23.333 \text{ ft/s}$$

Solution

Open an Excel spreadsheet, with column headings of t, v and v Δt. Fill the (A) column with t beginning at zero and increasing in steps of 0.1 until t = 2. Next calculate v in column B using $v = f(t) = t^3 + 2t^2 + 3t$ + 4 and the copy and paste command to fill in the column B for values of t between 0 and 2.0. In column C multiply the results in column B by 0.2 (the increment for Δx) for odd values of v. These odd values for t represent the mid points for the rectangles (slices) 0.2 units wide. In column C, we have calculated the areas of 10 rectangles each 0.2 units wide. Finally, we sum the results in column C and obtain the total area as 23.30 units. Performing the mathematical integration yields the area as 23.333. Our numerical integration was a bit low at 0.0333 units, which represents and error of 0.143%. Accuracy would be improved by increasing the number of slices, but most engineering calculations with errors less than a few percent are adequate.

t	v	vΔt
0	4	
0.1	4.321	0.864
0.2	4.688	
0.3	5.107	1.021
0.4	5.584	
0.5	6.125	1.225

0.6	6.736	
0.7	7.423	1.485
0.8	8.192	
0.9	9.049	1.810
1.0	10.000	
1.1	11.051	2.210
1.2	12.208	
1.3	13.477	2.695
1.4	14.864	
1.5	16.375	3.275
1.6	18.016	
1.7	19.793	3.959
1.8	21.712	
1.9	23.779	4.756
2.0	26.000	
	Area	23.300

Many students prefer to perform mathematical differentiation and integration, because it is less time consuming. However, engineers are rarely provided with mathematical expressions for displacement, velocity and acceleration. Instead, engineers conduct experiments and measure those quantities with respect to time. As a result, analysts usually work with digital records of measurements and perform numerical differentiation and integration in performing data analysis.

Integration of experimental data usually can be performed without introducing significant errors, as the integration process tends to smooth out the jitters in data produced by electronic noise. However, this is not the case of differentiation, which tends to amplify error due to electronic noise. The usual rule with differentiation of experimental data is to perform one differentiation with care and never try to differentiate the same data set twice.

2.8 SUMMARY

Kinematics deals with the motion of bodies without regard to the forces that are required to produce that motion. In this chapter we considered the motion of particles, which are small entities that may have mass but not physical dimensions. A particle is infinitesimally small when compared to distance it may travel.

Kinematics involves establishing the relations for the displacement, velocity and acceleration of a particle in motion using the time and geometry of the path of the particle. Forces required to produce this motion are not considered, although significant forces develop in order to produce the accelerations involved in certain problems.

When a particle moves in rectilinear motion, the position, velocity or acceleration are determined from:

$$v = \frac{dx}{dt} \qquad (2.2)$$

$$a = \frac{dv}{dt} = \frac{d^2x}{dt^2} \qquad (2.6)$$

$$a = v\frac{dv}{dx} \qquad (2.9)$$

$$\int_{x_0}^{x} dx = \int_{0}^{t}\left[v_0 + \int_{0}^{t} f(t)dt \right] dt \qquad (2.11)$$

If a particle is subjected to a constant acceleration a_0, then:

$$v = v_0 + a_0\, t \qquad (2.12)$$

$$x = x_0 + v_0\, t + a_0\, t^2/2 \qquad (2.13)$$

$$v^2 = v_0^2 + 2a_0(x - x_0) \qquad (2.14)$$

If a particle is traveling at a constant velocity,

$$a_0 = 0 \qquad \text{and} \qquad x = x_0 + v_c\, t \qquad (2.15)$$

If the acceleration of a particle is a known function of position given by $a = f(x)$, then:

$$\frac{1}{2}\left(v^2 - v_0^2\right) = \int_{x_0}^{x} f(x)dx \qquad (2.16)$$

If two particles P and Q are both moving along the x axis, their positions are given by x_P and x_Q respectively. The distance between the two particles is $x_{Q/P}$ is:

$$x_{Q/P} = x_Q - x_P \qquad (2.18a)$$

The velocity $v_{Q/P}$ is:

$$v_{Q/P} = v_Q - v_P \qquad (2.19a)$$

The acceleration $a_{Q/P}$ is:

$$a_{Q/P} = a_Q - a_P \qquad (2.20a)$$

The method for numerical differentiation employs differences Δy and Δt as:

$$\Delta y = y_2 - y_1 \qquad (2.21)$$

and

$$\Delta t = t_2 - t_1 \qquad (2.22)$$

then

$$\frac{\Delta x}{\Delta t} = \frac{x_2 - x_1}{t_2 - t_1}$$
(2.23)

The time corresponding to this derivative is given by:

$$t = (t_1 + t_2)/2$$
(2.24)

A numerical method for integration involves summation of small rectangular areas A_n of the total area A under a specified curve.

$$A = \sum_{n=1}^{N} A_n$$
(2.25)

CHAPTER 3

MOTION OF A PARTICLE IN THREE DIMENSIONS

INTRODUCTION

In the last chapter, we presented the kinematic equations that are used when dealing with a particle moving in a straight line (rectilinear motion). Unfortunately many particles and rigid bodies move along a curvilinear path. That fact implies that the particle change both speed and direction. With curvilinear motion, the coordinate systems we employ are more complicated. We still use Cartesian coordinates (x-y-z), but in many cases this choice of coordinates complicates the problem. Often it is simpler to use a different coordinate system. One common coordinate system that is used is the normal and tangential system as shown below:

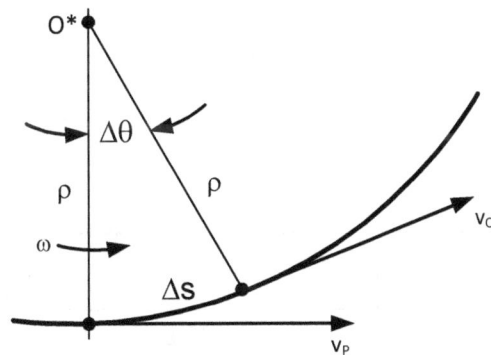

In this case the particle is traveling along a circular path with fixed radius ρ. We measure the distance to a point on the curve from a fixed point normal to the curvilinear path (ρ), the arc length (Δs), and an angle from a fixed line $\Delta\Theta$. This arrangement is referred to as a normal-tangential coordinate system with the third direction z that is perpendicular to the plane of the paper.

Another system that is commonly used is an $r - T(\Theta)$ system illustrated to the right:

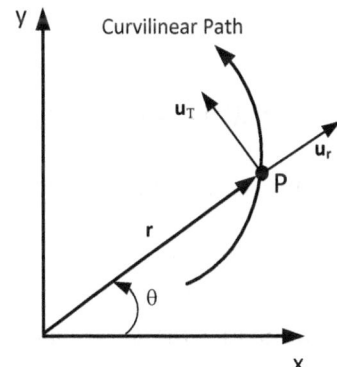

With this coordinate system, we measure the distance r from a fixed point to the curve and the angle θ from a fixed line to identify the point on the curve. This system is called a radial-transverse coordinate system. Once again, the third direction z is perpendicular to the plane of the paper.

We mainly deal with motion in a plane identified with either the x-y, the n-t (normal - tangential), or the r-T (Θ) (radial/transverse) coordinate systems. The remainder of this chapter will be devoted to developing kinematic equations for particle motion in these three coordinate systems. In Chapter 4, we will develop the kinematic equations in these three coordinate systems for rigid bodies undergoing planar motion.

CONCEPT PROBLEM

You are testing the accuracy of a new low-velocity launch system. The system is to launch a package that reaches a target that is initially 500 m from the launcher and is moving away at a velocity of 25 m/s. The launcher is aligned with the road and the road is level. If the launcher velocity is set at 100 m/s, determine the angle of the launch tube so that the package reaches the target. Also determine the flight time required.

To begin let's prepare a drawing showing the position of the target.

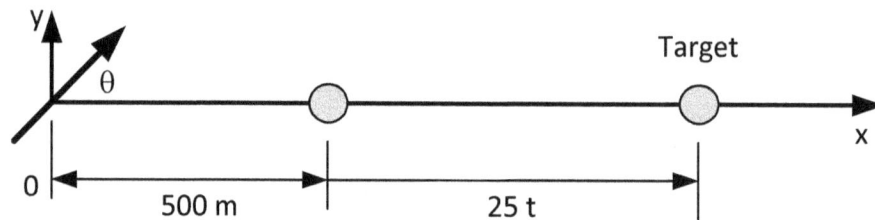

The target is moving along the x axis. Let's write an equation that locates its position x_T with respect to time.

$$x_T = 500 + 25\ t \tag{a}$$

The package is moving along the x axis as well, although it is many meters above the ground level. The horizontal position of the package is given by:

$$x_P = (v \cos \theta)\ t \tag{b}$$

where v = 100 m/s is the launch velocity.

At impact of the package with the target:

$$x_T = x_P \tag{c}$$

Substitute Eqs. (a) and (b) into Eq. (c) to obtain the time of impact as:

$$t_I = 500/(v \cos \theta - 25) \qquad (d)$$

The time t_I is when the target and the package are at the same position along the x axis; however, at this time the package may be several meters above the target or buried in the roadway.

A second requirement is for the package to be at ground level (y = 0) at the impact time. To determine this time, let's write a relation for the height of the package with respect to time using Eq. (2.13):

$$y_P = (v \sin \theta) t - \tfrac{1}{2} g t^2 = 0 \qquad (e)$$

or

$$t_I = 2v \sin \theta/g \qquad (f)$$

Equating Eqs. (d) and (f) yields

$$v^2 \sin \theta \cos \theta - 25 v \sin \theta = 250 g = 2,452$$

$$v \sin \theta(v \cos \theta - 25) = 2,452 \qquad (g)$$

We employed a spreadsheet to evaluate $v \sin \theta(v \cos \theta - 25)$ as a function of θ with v = 100 m/s.

theta	theta, rad	v sin theta	v cos theta	v cos theta - 25	Third Col times fifth Col
0	0	0	100	75	0
5	0.0873	8.72	99.62	74.62	650.33
10	0.1745	17.36	98.48	73.48	1275.94
15	0.2618	25.88	96.59	71.59	1852.90
20	0.3491	34.20	93.97	68.97	2358.83
21	0.3665	35.84	93.36	68.36	2449.68
21.03	0.3670	35.88	93.34	68.34	2452.35
21.1	0.3683	36.00	93.30	68.30	2458.56
25	0.4363	42.26	90.63	65.63	2773.62
30	0.5236	50.00	86.60	61.60	3080.08
35	0.6108	57.36	81.92	56.92	3264.50
40	0.6981	64.28	76.61	51.61	3317.07
45	0.7854	70.71	70.71	45.71	3232.27
50	0.8726	76.60	64.28	39.28	3009.01
55	0.9599	81.91	57.36	32.36	2650.72
60	1.0472	86.60	50.00	25.00	2165.26

The results from the spreadsheet show that:

$$\theta = 21.03 \text{ degrees}$$

The impact time is:

$$t_I = 2v \sin \theta/g = 2(100) \sin (21.03^\circ)/9.807 = 7.318 \text{ s} \qquad (h)$$

DISCUSSION

Most gun-projectile applications are military, where the purpose is to eliminate enemy installations, equipment and personnel or reduce the threat that they represent. The gun-projectile combination depends on the target. For example a mortar usually involves small-diameter, portable gun tubes that are sufficiently light to be carried by a small team into a battle area. The projectiles, 81 mm in diameter, are loaded in the muzzle and drop to the base of the tube, where they initiate a charge that fires the projectile with an exit velocity of about 700 ft/s. The range is determined by the angle of the tube, which can be quickly adjusted. The range is inversely proportional to the angle of inclination of the mortar tube. The targets for mortars are enemy personnel and light vehicles.

On the other hand, North Korea's 170-millimeter Koksan self-propelled gun has a range of 40 km with rocket boosters. Large projectiles, such as those delivered by this gun, are to destroy major infrastructure like bridges, railway terminals and buildings.

Korea's 170-millimeter Koksan self-propelled gun.

Projectiles differ as well depending on the targets involved. To destroy heavily armored tanks there are two approaches. One is to use kinetic energy projectile without a warhead but with high density. These projectiles are made of tungsten or depleted uranium for high density (providing a high mass) and fired out of low caliber gun tubes at very high velocity to maximize the kinetic energy of the projectile to a value sufficient to penetrate thick armor. Another approach is to use shaped charges, which when detonated, develop a liquid jet of metal at high velocity that penetrates the tanks armor.

Still another approach for destroying personnel is the employ proximity fuzing on the projectile. The proximity fuze is set to detonate the projectile some distance above ground level (15 to 20 ft). The resulting detonation fragments the projectile sending fragments that cover a large circular area inflicting maximum damage on personnel in this area.

Mortars are also used for fireworks displays. These mortars range in size up to a foot in diameter and fire pyrotechnic displays several hundred feet into the air, where the projectile detonates creating an interesting visual display. An illustration, with tubes used to launch firework projectiles that have diameters ranging from 2 to 8 in. in diameter, are shown below.

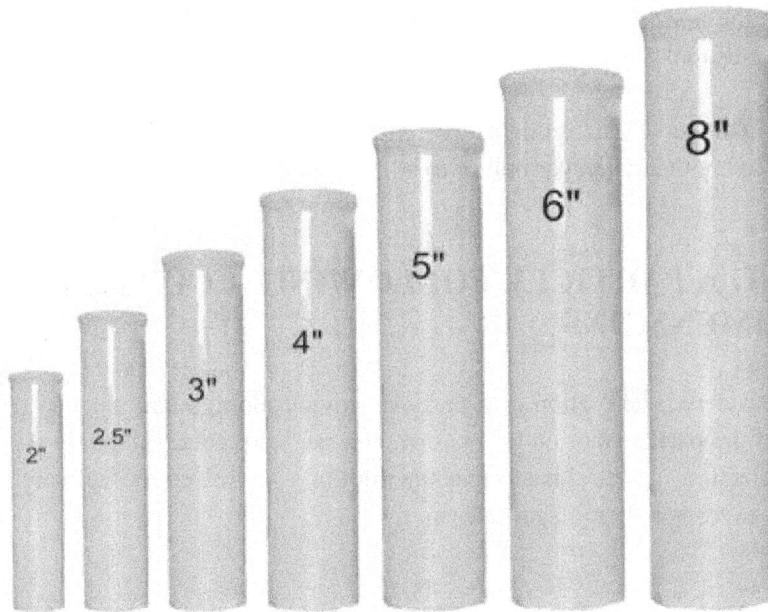

MORE DISCUSSION

There many more applications of projectile motion that do not involve guns or projectiles.

In sports, a basketball player shooting for 3 points, launches the basketball with a certain velocity and angle so that the ball passes through the hoop and engages the net. In football, the place kicker kicks the ball with an angle sufficiently large to avoid the outstretched hands of the opposing players with sufficient velocity to insure the range necessary for the ball to pass through the uprights. In baseball the pitcher delivers the ball with angle and velocity to pass through the strike zone. The pitcher also generates spin on the ball as it leaves his or her hand. This spin creates an aerodynamically induced force that is not accounted for in the traditional analysis of projectile motion. The spin cause the ball to curve or to drop or hop so as to cause the batter who may be expecting the normal trajectory of the ball to swing and miss.

The jump in skiing is another example of projectile motion. The skier gains velocity in a downhill track that ends in a ramp with a slight upward angle. The skier is launched into the air and lands on a downhill slope. Poland's Kamil Stoch successfully defended his large hill crown on February 17, 2018. In this event he landed safely with a 136.5 m jump. A photo of Kamil in midair during this jump is show to the right.

Liquid ejecting from the nozzle on the end of a pressurized hose can be analyzed using the relations for projectile motion. An example is a fireman delivering water to select areas of a burning building.

Conveyer belts may also serve as a launch tube, with velocities are usually small. The projectile is a carton that is moved to another conveyer belt or to a loading dock.

3.1 TRACKING A PARTICLE FOLLOWING A CURVILINEAR PATH IN THREE DIMENSIONS

In Chapter 2, we considered the motion of a particle moving along a straight line (i.e. the x axis). As such, the direction of the particle motion was known and we were concerned only with the magnitude of its velocity and acceleration. With this straight line motion, we treated the velocity and acceleration as scalar quantities and expressed their magnitudes as:

$$v = \frac{dx}{dt} \tag{2.2}$$

and

$$a = \frac{dv}{dt} = \frac{d^2x}{dt^2} \tag{2.6}$$

If the position x is known as a function of time t, it is easy to differentiate Eqs. (2.2) and (2.6) to obtain the magnitude of the velocity and the acceleration of the particle as it moves along the x axis. However, for a particle moving along a curved path, the velocity and acceleration are both vector quantities, because their magnitude and direction both change with time. To examine more closely the curvilinear motion of a particle, consider the curved path shown in Fig. 3.1.

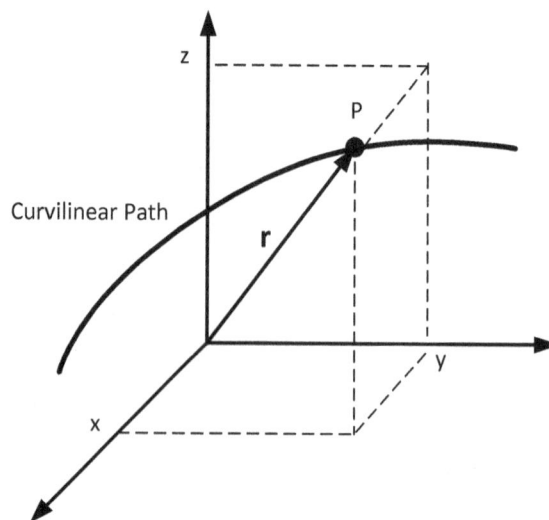

Fig. 3.1 A particle P moves along a curvilinear path with its location identified by a position vector **r**.

In developing the equations describing the velocity and acceleration of the particle as it moves along the curved path, we will reference the motion to three different coordinate systems: rectangular coordinates, normal and tangential coordinates and radial and transverse coordinates. The three different coordinate systems are provided to enable you to more effectively determine velocity and acceleration components for particles moving along paths with different curvilinear geometries.

In **rectangular** coordinates, the equation for the position vector **r** is given by:

$$\mathbf{r} = x\mathbf{i} + y\mathbf{j} + z\mathbf{k} \tag{3.1}$$

where **i**, **j** and **k** are unit vectors in the x, y and z directions respectively.

The velocity vector **v** is determined by differentiating the position vector with respect to time to yield:

$$\mathbf{v} = dx/dt\ \mathbf{i} + dy/dt\ \mathbf{j} + dz/dt\ \mathbf{k} \tag{3.2}$$

where $\quad v_x = dx/dt \qquad\qquad v_y = dy/dt \qquad\qquad v_z = dz/dt$

then $\qquad\qquad\qquad \mathbf{v} = v_x\ \mathbf{i} + v_y\ \mathbf{j} + v_z\ \mathbf{k} \tag{3.3}$

The magnitude of the velocity vector is given by:

$$v = [v_x^2 + v_y^2 + v_z^2]^{1/2} \tag{3.4}$$

The direction of the velocity vector is expressed in terms of the unit vector $\mathbf{u_v}$ as:

$$\mathbf{u_v} = \mathbf{v}/v \tag{3.5}$$

$\mathbf{u_v}$ is a vector in the same direction as the velocity but it has a magnitude of one.

The acceleration of the particle, is determined by differentiating the velocity vector with respect to time to give:

$$\mathbf{a} = d^2x/dt^2\ \mathbf{i} + d^2y/dt^2\ \mathbf{j} + d^2z/dt^2\ \mathbf{k} \tag{3.6}$$

where $\quad a_x = d^2x/dt^2 \qquad\qquad a_y = d^2y/dt^2 \qquad\qquad a_z = d^2z/dt^2$

then $\qquad\qquad\qquad \mathbf{a} = a_x\ \mathbf{i} + a_y\ \mathbf{j} + a_z\ \mathbf{k} \tag{3.7}$

The magnitude of the acceleration vector is given by:

$$a = [a_x^2 + a_y^2 + a_z^2]^{1/2} \tag{3.8}$$

The direction of the acceleration vector is expressed in terms of a unit vector $\mathbf{u_a}$ as:

$$\mathbf{u_a} = \mathbf{a}/a \tag{3.9}$$

3.1.1 Planar Particle Motion

If we restrict the particle motion to the x-y plane, where z = 0, then the relations for the position, velocity and acceleration vectors are obtained by differentiating the position vector. The equation for the position vector $\mathbf{r}$ is given by:

$$\mathbf{r} = x\mathbf{i} + y\mathbf{j} \tag{3.10}$$

The velocity vector v is:

$$\mathbf{v} = dx/dt\ \mathbf{i} + dy/dt\ \mathbf{j} \tag{3.11}$$

where $\qquad v_x = dx/dt \qquad\qquad v_y = dy/dt$

$$\mathbf{v} = v_x\,\mathbf{i} + v_y\,\mathbf{j} \tag{3.12}$$

The magnitude of the velocity vector is given by:

$$v = [v_x^2 + v_y^2]^{1/2} \tag{3.13}$$

The unit vector and the tangent angle θ are given by:

$$\mathbf{u_v} = \mathbf{v}/v \qquad \text{or} \qquad \theta = \tan^{-1} v_y/v_x \tag{3.14}$$

where θ is the angle that the velocity vector $\mathbf{v}$ makes with the x axis.

The acceleration of the particle is given by:

$$\mathbf{a} = d^2x/dt^2\ \mathbf{i} + d^2y/dt^2\ \mathbf{j} \tag{3.15}$$

where $\qquad a_x = d^2x/dt^2 \qquad\qquad a_y = d^2y/dt^2$

$$\mathbf{a} = a_x\,\mathbf{i} + a_y\,\mathbf{j} \tag{3.16}$$

The magnitude of the acceleration vector is given by:

$$a = [a_x^2 + a_y^2]^{1/2} \tag{3.17}$$

The tangent angle θ of the acceleration vector is:

$$\boldsymbol{\theta} = \tan^{-1} a_y/a_x \tag{3.18}$$

EXAMPLE 3.1

A particle moves along a curvilinear path defined by $y^2 = 9x$, with its position relative to the x axis given by $x = 4t^2 + 1$. Positions x and y are expressed in m, time in s and velocity in m/s. Determine the equations for the magnitude and direction of the velocity vector as a function of time.

Solution:

Let's begin by differentiating x with respect to t:

$$x = 4t^2 + 1 \qquad \text{and} \qquad v_x = dx/dt = 8t \qquad (a)$$

Note that $y = \pm 3\, x^{1/2}$. Selecting the positive root, permits us to write: $y = 3\,(4t^2 + 1)^{1/2}$ (b)

Differentiating the expression for y in Eq. (b) gives:

$$v_y = \frac{dy}{dt} = \frac{3}{2}\frac{8t}{\sqrt{4t^2+1}} = \frac{12t}{\sqrt{4t^2+1}} \qquad (c)$$

The magnitude of the velocity vector is given by substituting Eqs. (a) and (c) into Eq. (3.13) to obtain:

$$v = \sqrt{v_x^{\,2} + v_y^{\,2}} = \sqrt{(8t)^2 + \left[\frac{12t}{\sqrt{4t^2+1}}\right]^2} \qquad (d)$$

and the direction of the velocity vector is given by Eq. (3.14) as:

$$\tan\theta = \frac{v_y}{v_x} = \frac{12t}{8t\sqrt{4t^2+1}} = \frac{3}{2\sqrt{4t^2+1}} \qquad (e)$$

where θ is the angle that the velocity vector **v** makes with the x axis.

Let $t = 2$ s and determine the magnitude of the velocity components as:

$$v_x = dx/dt = 8t = 8(2) = 16 \text{ m/s}$$

$$v_y = \frac{dy}{dt} = \frac{12t}{\sqrt{4t^2+1}} = \frac{12(2)}{\sqrt{4(2)^2+1}} = \frac{24}{\sqrt{17}} = 5.821\,\text{m/s} \qquad (f)$$

The magnitude of the velocity vector is given by

$$v = \sqrt{v_x^{\,2} + v_y^{\,2}} = \sqrt{(16)^2 + (5.821)^2} = 17.03\,\text{m/s} \qquad (g)$$

and the direction of the velocity vector is given by:

$$\tan\theta = \frac{v_y}{v_x} = \frac{5.821}{16} = 0.3638 \qquad \text{and} \qquad \theta = 19.99 \text{ degrees} \qquad (h)$$

EXAMPLE 3.2

A particle in planar motion follows a curved path described by coordinates:

$$x = t + 1/t = (t^2 + 1)/t \qquad \text{and} \qquad y = t - 1/t = (t^2 - 1)/t \qquad \text{(a)}$$

where x and y are measured in ft and time in s.

Determine the equation for the path of the particle and its velocity. Also determine the magnitude of the velocity and its direction when t = 2 s.

Solution:

To determine the path of a particle when given the equations for the positions x and y as a function of time, we seek expressions for x and y that enable us to eliminate the variable t. This often involves a lot of trial and error until we find the correct functions. In this example, we begin by squaring the expressions for the position parameters x and y to give:

$$x^2 = \frac{t^4 + 2t^2 + 1}{t^2} \qquad\qquad y^2 = \frac{t^4 - 2t^2 + 1}{t^2} \qquad\qquad \text{(b)}$$

Subtract y^2 from x^2 to obtain the expression for the path of the particle, which is independent of time:

$$x^2 - y^2 = \frac{4t^2}{t^2} = 4 \qquad\qquad \text{(c)}$$

Determine the components of the velocity by differentiating Eq. (a) to obtain:

$$v_x = \frac{dx}{dt} = \frac{t(2t) - (t^2 + 1)(1)}{t^2} = \frac{t^2 - 1}{t^2}$$

$$v_y = \frac{dy}{dt} = \frac{t(2t) - (t^2 - 1)(1)}{t^2} = \frac{t^2 + 1}{t^2} \qquad\qquad \text{(d)}$$

The magnitude of the velocity is given by:

$$v = \sqrt{v_x^2 + v_y^2} = \sqrt{\frac{t^4 - 2t^2 + 1}{t^4} + \frac{t^4 + 2t^2 + 1}{t^4}} = \sqrt{2\left(\frac{t^4 + 1}{t^4}\right)} \qquad\qquad \text{(e)}$$

and its direction by:

$$\tan\theta = \frac{v_y}{v_x} = \frac{(t^2 + 1)t^2}{t^2(t^2 - 1)} = \frac{t^2 + 1}{t^2 - 1} \qquad\qquad \text{(f)}$$

For t = 2 s, we find:

$$v = \sqrt{2\left(\frac{t^4 + 1}{t^4}\right)} = \sqrt{2\left(\frac{(2)^4 + 1}{(2)^4}\right)} = \sqrt{\frac{17}{8}} = 1.458 \, m/s \tag{g}$$

and the direction of the velocity vector at t = 2 s is:

$$\tan\theta = \frac{v_y}{v_x} = \frac{t^2 + 1}{t^2 - 1} = \frac{5}{3} = 1.667 \quad \text{and} \quad \theta = 59.04 \text{ degrees} \tag{h}$$

EXAMPLE 3.3

A particle, initially at the origin with x = y = z = 0 when t = 0, moves along a curvilinear path with a velocity given by:

$$v = 6\,t^2\,\mathbf{i} - 8\,t\,\mathbf{j} + 4\,\mathbf{k} \tag{a}$$

where v is given in units of ft/s.

Determine the position of the particle at t = 2 s, the acceleration at t = 3 s and the equation of the path of the particle.

Solution:

The position:

$v_x = dx/dt = 6\,t^2$ Integrate to give $x = 6t^3/3 + C_1 = 2t^3$ because $C_1 = 0$ (b)

$v_y = dy/dt = -8\,t$ Integrate to give $y = -8t^2/2 + C_2 = -4t^2$ because $C_2 = 0$ (c)

$v_z = dx/dt = 4$ Integrate to give $z = 4t + C_3 = 4t$ because $C_3 = 0$ (d)

at t = 2 s,
$$x = 2(2)^3 = 16 \text{ ft} \qquad y = -4(2)^2 = -16 \text{ ft} \qquad z = 4(2) = 8 \text{ ft} \tag{e}$$

Then the position vector at t = 2 s is:

$$\mathbf{r} = 16\,\mathbf{i} - 16\,\mathbf{j} + 8\,\mathbf{k} \tag{f}$$

The acceleration:

Differentiating Eq. (b) with respect to t gives:

$$a_x = d^2x/dt^2 = 12\,t \tag{g}$$

Differentiating Eq. (c) with respect to t gives:

$$a_y = d^2y/dt^2 = -8 \qquad\qquad\qquad \text{(h)}$$

Differentiating Eq. (d) with respect to t gives:

$$a_z = d^2y/dt^2 = 0 \qquad\qquad\qquad \text{(i)}$$

at t = 3 s

$$a_x = 36 \text{ ft/s}^2 \qquad a_y = -8 \text{ ft/s}^2 \qquad a_z = 0 \qquad \text{(j)}$$

and

$$a = [(36)^2 + (-8)^2]^{1/2} = 36.88 \text{ ft/s}^2 \qquad\qquad \text{(k)}$$

$$\mathbf{a} = 36\,\mathbf{i} - 8\,\mathbf{j} \qquad\qquad\qquad \text{(l)}$$

The equation for the path of the particle is determined from the position coordinates given in Eqs. (b), (c) and (d). We solve for t as a function of x and substitute this result into the expressions for y and z to obtain:

$$x = 2t^3 \qquad y = -4t^2 \qquad z = 4t \qquad\qquad \text{(m)}$$

Note: $\quad t = (x/2)^{1/3} \quad$ then $\quad y = -4(x/2)^{2/3} \qquad z = 4(x/2)^{1/3} \qquad \text{(o)}$

After eliminating t from y and z add y and z^2 to give the equation of the path of the particle:

$$y + z^2 = -4(x/2)^{2/3} + 16(x/2)^{2/3} = 12(x/2)^{2/3} \qquad\qquad \text{(p)}$$

3.2 PROJECTILE MOTION

Projectiles are often fired from guns, but vehicles traveling over a speed bump, water flowing from the end of a hose or a skier competing in a jump event may all be modeled as a projectile. However, we must know the tangent of the curve, along which the projectile is moving, as it exits into free space . In all these situations, we model the object (projectile) as a particle initially moving along a straight line at some velocity and an angle relative to the ground plane. At some point, the particle is released from its constraint and when it becomes free, it is subjected to gravitational acceleration. For example, when a projectile exits a gun tube, it has an exit velocity and a direction determined by the orientation of the gun tube. Immediately upon exiting the gun tube, the projectile is in free flight and is subjected to gravitational acceleration[1].

Let's consider a projectile, as it exits the gun tube with a velocity of v_{ex}. If the gun tube lies in the x-y plane and makes an angle θ with the x axis, the initial velocities in the x and y directions are given by:

$$v_x)_0 = v_{ex} \cos\theta \qquad\qquad\qquad \text{(3.19a)}$$

$$v_y)_0 = v_{ex} \sin\theta \qquad\qquad\qquad \text{(3.19b)}$$

---

[1] The projectile in the gun tube is subject to the gravitational force but it is small in comparison to the force due to the pressure of the explosively generated gasses and is neglected in these analyses.

After launch the projectile is subjected to the gravitational acceleration g, where $g = 9.81 \text{ m/s}^2$. The effect of gravity on the projectile is to reduce its velocity in the y direction with time by 9.81 m/s². We express this decrease in velocity as a linear function of t as given by:

$$v_y = v_y)_0 - gt \qquad (3.20)$$

Let's integrate Eq. (3.20) to obtain a relation for y as a function of time:

$$y = v_y)_0 t - gt^2/2 + y_0 \qquad (3.21)$$

where y_0, the integration constant represents the elevation of the ground, upon which the gun tube is located. If the gun tube is initially located on the ground plane, it is evident that $y_0 = 0$ and Eq. (3.21) becomes:

$$y = v_y)_0 t - gt^2/2 \qquad (3.22)$$

By setting y = 0 in Eq. (3.22), we can solve for t and determine the time of the flight of the projectile before it impacts the ground plane.

$$t_{Impact} = \frac{2v_y)_0}{g} \qquad (3.23)$$

If the ground plane is level, Eq. (3.23) is valid and the range x_R is given by:

$$x_R = v_x)_0 \, t_{Impact} \qquad (3.24)$$

EXAMPLE 3.4

A projectile is fired vertically into the air at an initial velocity of 300 m/s. Determine the maximum height that the projectile achieves and the time when it strikes the ground. Neglect friction effects due to air resistance.

Solution:

The height of the projectile is given by Eq. (3.20) as:

$$y = v_0 t - \tfrac{1}{2} gt^2 \qquad (a)$$

If we differentiate Eq. (a) with respect to time, we obtain the velocity component v_y in the vertical direction. At the maximum height (the apogee) of the projectile, the velocity is zero; hence, we can write:

$$v_y = \frac{dy}{dt} = v_0 - gt = 0 \qquad (b)$$

Setting $v_y = 0$ recognizes that the y component of the velocity vanishes at the highest point in the projectile's trajectory.

Solving Eq. (b) gives the time t for the projectile to reach its maximum height:

$$t = v_0/g = 300/9.81 = 30.58 \text{ s} \qquad (c)$$

Substituting the results for the time t = 30.58 s into Eq. (a) gives the maximum height as:

$$y_{Max} = 300(30.58) - \tfrac{1}{2}(9.81)(30.58)^2 = 4{,}587 \text{ m} \qquad \text{(d)}$$

To determine the time to strike the ground, set y in Eq. (a) equal to zero and solve for the time t as:

$$y = v_0 t - \tfrac{1}{2} gt^2 = 0 \qquad \text{(e)}$$

Solving for t gives:

$$t = 2v_0/g = 2(300)/9.81 = 61.16 \text{ s} \qquad \text{(f)}$$

EXAMPLE 3.5

A mortar is used to lob a projectile high in the air, but with a limited range. A projectile exits a mortar tube at an angle of 80° with a muzzle velocity $v_m = 120$ m/s, as illustrated in Fig. E.3.5. Determine the range of the projectile.

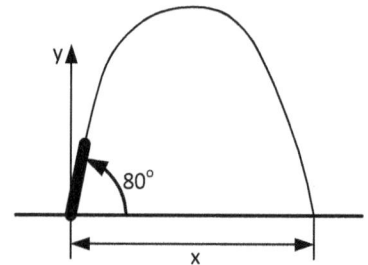

Fig. E3.5 Projectile's path exiting a mortar tube at 80°.

Solution:

The initial velocities of the projectile in the x and y directions are:

$$v_x)_0 = v_m \cos 80^O = 120 \cos 80^O = 20.84 \text{ m/s} \qquad \text{(a)}$$

$$v_y)_0 = v_m \sin 80^O = 120 \sin 80^O = 118.2 \text{ m/s} \qquad \text{(b)}$$

The velocity in the y direction as a function of time t is:

$$v_y = v_y)_0 - gt = 118.2 - 9.81t \qquad \text{(c)}$$

where $g = 9.81$ m/s^2 is the gravitational acceleration.

Integrating Eq. (c) gives:

$$y = 118.2\, t - 9.81\, t^2/2 \qquad \text{(d)}$$

Because the mortar is located on the ground plane, the constant of integration is zero in Eq. (d). Setting y = 0 in Eq. (d) and solving for time t gives the time to target of t = 24.10 s.

The projectile impacts the ground plane with a range of:

$$x = v_x)_0\, t = 20.84(24.10) = 502.2 \text{ m} \qquad \text{(e)}$$

EXAMPLE 3.6

A ball is thrown vertically upward with an initial velocity of 30 ft/s. One second later a second ball is thrown vertically upward with an initial velocity of 25 ft/s. It is evident that the second ball will not travel as high as the first ball. Both balls will be at the same height sometime during the period when they are both falling downward. However, the first ball could be moving downward while the second ball is still going up. Determine the time when the two balls will be an equal distance from the ground. Also determine the height of both balls at this time.

Solution:

The height of ball #1 is given by:
$$y_1 = v_0)_1\, t - \tfrac{1}{2}\, g t^2 \qquad\qquad (a)$$

The height of ball #2 is:
$$y_2 = v_0)_2\, (t - 1) - \tfrac{1}{2}\, g(t - 1)^2 \qquad\qquad (b)$$

The balls are the same height from the ground when $y_1 = y_2$: hence:

$$v_0)_1\, t - \tfrac{1}{2}\, g t^2 = v_0)_2\, (t - 1) - \tfrac{1}{2}\, g(t^2 - 2t + 1) \qquad\qquad (c)$$

Substituting numerical values into Eq. (c) yields:

$$30t = 25(t - 1) + 32.17t - \tfrac{1}{2}\,(32.17) \qquad\qquad (d)$$

Solving Eq. (d) for the time t yields:
$$t = 1.512\ \text{s} \qquad\qquad (e)$$

Substituting for t into Eq. (a) gives:

$$y_1 = v_0)_1\, t - \tfrac{1}{2}\, g t^2 = 30(1.512) - (32.17/2)(1.512)^2 = 8.585\ \text{ft}$$

$$\qquad\qquad (f)$$

$$y_2 = v_0)_2\, (t - 1) - \tfrac{1}{2}\, g(t - 1)^2 = 25(0.512) - (32.17/2)(0.512)^2 = 8.585\ \text{ft}$$

EXAMPLE 3.7

A cannon fires a projectile with an exit velocity $v_{ex} = 1{,}600$ ft/s, as shown in Fig. E3.7. The angle of the cannon relative to ground level is 35°. If we assume the ground is level over the range of the projectile and that the air resistance against the projectile can be neglected, determine the maximum height of the projectile and the distance it travels before impacting the ground.

Fig. E3.7 A cannon firing a projectile at a 35° inclination to the ground plane.

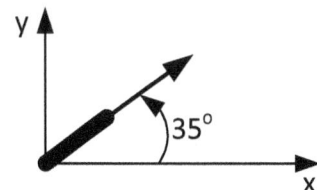

Solution:

Determine the components of the exit velocity:

$$(v_y)_0 = v_{ex} \sin \theta = 1{,}600 \sin 35^O = 917.7 \text{ ft/s} \qquad (a)$$

$$(v_x)_0 = v_{ex} \cos \theta = 1{,}600 \cos 35^O = 1{,}311 \text{ ft/s} \qquad (b)$$

The velocity component v_y is given by:

$$v_y = (v_y)_0 - gt = 917.7 - 32.17\, t \qquad (c)$$

Integrating Eq. (c) yields:

$$y = 917.7\, t - 32.17\, t^2 /2 \qquad (d)$$

Because $(v_y)_0 = 0, \quad y = 0$ at $t = 0$

The distance the projectile travels in the x direction is given by integrating $(v_x)_0$ with respect to time.

$$x = (v_x)_0\, t = 1{,}311\, t \qquad (e)$$

Because $(v_x)_0 = 0. \quad x = 0$ at $t = 0$

The projectile impacts the ground when $y = 0$; hence, we use Eq. (d) and write:

$$y = 917.7\, t - 32.17\, t^2/2 \; = 0 \qquad (f)$$

Solving Eq. (f) for t gives

$$t = \frac{2(917.7)}{32.17} = 57.05 \text{ s} \qquad (g)$$

Substituting the result for the time of impact into Eq. (e) gives the distance traveled by the projectile as:

$$x = (1{,}311)(57.05) = 74{,}800 \text{ ft or } 14.17 \text{ mile} \qquad (h)$$

The maximum height of the projectile occurs when $v_y = 0$; hence, we employ Eq. (c) to determine the time of the maximum elevation as:

$$v_y = 917.7 - 32.17\, t = 0 \qquad \text{which gives } t = 28.53 \text{ s} \qquad (i)$$

Substituting this value for time t into Eq. (d) gives:

$$y = 917.7\,(28.53) - 32.17\,(28.53)^2 /2$$

$$y = 26{,}182 - 13{,}093 = 13{,}089 \text{ ft or } 2.479 \text{ mile} \qquad (j)$$

Examining the results indicates that artillery capable of launching projectiles with relatively high exit velocities have very long range and can inflict damage on targets located many miles from the artillery emplacement.

EXAMPLE 3.8

A baseball pitcher throws a straight pitch with a velocity of 84 MPH. When the ball leaves the pitcher's hand, it is 6 ft above ground level and makes an angle of 5° with the horizontal plane, as shown in Fig. E3.8. At what level does the ball cross the plate, when it reaches the batter?

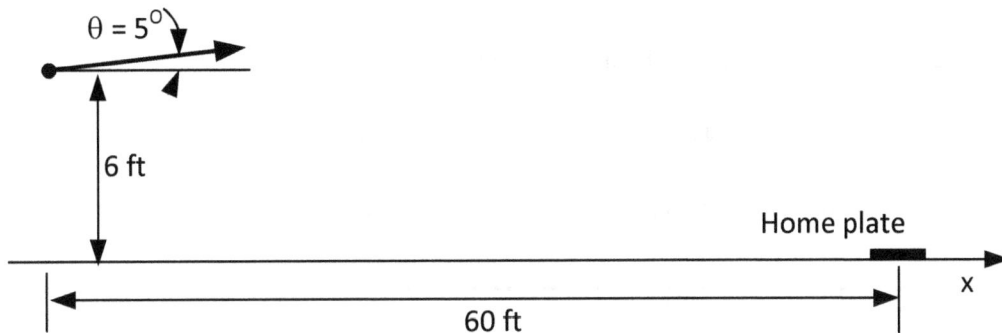

Fig. E3.8 A baseball pitch is thrown with a velocity of 84 MPH and an inclination of 5°.

Solution:

First let's convert 84 MPH into units of ft/s by:

$$v = \frac{84 \, \text{mile}}{\text{hour}} \times \frac{5280 \text{ft}}{\text{mile}} \times \frac{\text{hour}}{3600 \text{s}} = 123.2 \, \text{ft} / \text{s} \tag{a}$$

The initial components of the velocity in the x and y directions are:

$$(v_x)_0 = v \cos \theta = 123.2 \cos 5^\circ = 122.7 \, \text{ft} / \text{s} \tag{b}$$

$$(v_y)_0 = v \sin \theta = 123.2 \sin 5^\circ = 10.74 \, \text{ft} / \text{s} \tag{c}$$

The time required for the ball to reach the plate is given by:

$$t = \frac{d}{(v_x)_0} = \frac{60}{122.7} = 0.4890 \, \text{s} \tag{d}$$

The expression for v_y as a function of time is given by:

$$v_y = (v_y)_0 - gt = 10.74 - 32.17 \, t \tag{e}$$

Then:

$$y_t = (v_y)_0\, t - gt^2/2 = 10.74(0.4890) - 32.17(0.4890)^2/2 = 5.252 - 3.846 = 1.406 \text{ ft} \qquad (f)$$

However, the baseball was at a height of 6 ft before it was released; hence, the pitch was at a height of:

$$y_{d=60} = (y)_0 + y_{t\,=\,0.4890} = 6 + 1.406 = 7.406 \text{ ft} \qquad (g)$$

Not quite a wild pitch but close!

EXAMPLE 3.9

A particle moves along a path described by its coordinates, which are:

$$x = 2t^2(t - 4) \qquad \text{and} \qquad y = 3t(t - 4)$$

Determine the velocity and acceleration after 2 s. The units are in. and s.

Solution

The components of velocity are determined by differentiating x and y with respect to time:

$$v_x = \frac{dx}{dt} = 2t^2 + (t-4)(4t) = 2t(3t-8) \qquad (a)$$

and

$$v_y = \frac{dy}{dt} = 6t - 12 = 6(t-2) \qquad (b)$$

At t = 2 s, the velocity components are:

$$v_x = 4(6-8) = -8 \text{ in}/\text{s} \quad \text{and} \qquad v_y = 0 \qquad (c)$$

The magnitude of the velocity is:

$$v = [\,v_x^2 + v_y^2\,]^{1/2} = [(-8)^2 + 0]^{1/2} = 8 \text{ in/s} \qquad (d)$$

The direction of the velocity vector with respect to the negative x axis is given by:

$$\theta = \tan^{-1}\frac{v_y}{v_x} = \tan^{-1}\frac{0}{-8} = 0° \qquad (e)$$

The components of acceleration are:

$$a_x = \frac{dv_x}{dt} = 4(3t - 4) \qquad \text{and} \qquad a_y = \frac{dv_y}{dt} = 6 \text{ in / s}^2 \qquad \text{(f)}$$

At $t = 2$ s the magnitude of the two acceleration components is:

$$a_x = 4(6 - 4) = 8 \text{ in/s}^2 \qquad \text{and} \qquad a_y = 6 \text{ in/s}^2 \qquad \text{(g)}$$

$$a = [a_x^2 + a_y^2]^{1/2} = [(8)^2 + (6)^2]^{1/2} = 10 \text{ in/s} \qquad \text{(h)}$$

The direction of the acceleration vector is given by:

$$\theta = \tan^{-1}\frac{a_y}{a_x} = \tan^{-1}\frac{6}{8} = 36.87° \qquad \text{(i)}$$

EXAMPLE 3.10

A particle moves along a path described by its coordinates that are:

$$x = 48 \sin (\pi t/2) \qquad \text{and} \qquad y = 15\,t^2$$

Determine the velocity and acceleration after 2 s. The units are meters and seconds.

Solution

The velocity components are determined by differentiating x and y with respect to time:

$$v_x = \frac{dx}{dt} = 48\left(\frac{\pi}{2}\right)\cos\left(\frac{\pi t}{2}\right) = 24\pi \cos\left(\frac{\pi t}{2}\right) \qquad v_y = \frac{dy}{dt} = 30t \qquad \text{(a)}$$

At $t = 2$ s the velocity components reduce to:

$$v_x = 24\pi \cos(\pi) = -24\pi \text{ m / s} \qquad \text{and} \qquad v_y = 60 \text{ m/s} \qquad \text{(b)}$$

The magnitude of the velocity is:

$$v = [v_x^2 + v_y^2]^{1/2} = [(24\pi)^2 + (60)^2]^{1/2} = 96.36 \text{ m/s} \qquad \text{(c)}$$

The direction of the velocity vector relative to the x axis is given by:

$$\theta = \tan^{-1}\frac{v_y}{v_x} = \tan^{-1}\frac{60}{-24\pi} = -38.51° \qquad \text{(d)}$$

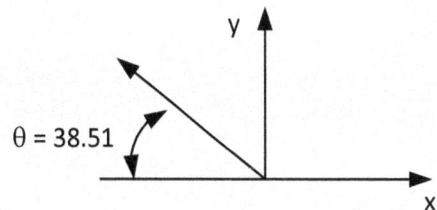

$\theta = 38.51$

The components of acceleration are:

$$a_x = \frac{dv_x}{dt} = -\left(24\pi\right)\left(\frac{\pi}{2}\right)\sin\left(\frac{\pi t}{2}\right) = -12\pi^2\sin\left(\frac{\pi t}{2}\right) \quad \text{and} \quad a_y = \frac{dv_y}{dt} = 30\,\text{m/s}^2 \quad (e)$$

At t = 2 s the magnitude of the acceleration is:

$$a = [a_x^2 + a_y^2]^{1/2} = [(0)^2 + (30)^2]^{1/2} = 30\,\text{m/s} \tag{f}$$

The direction of the acceleration vector is given by:

$$\theta = \tan^{-1}\frac{a_y}{a_x} = \tan^{-1}\frac{30}{0} = 90° \tag{g}$$

3.3 NORMAL AND TANGENTIAL COMPONENTS OF ACCELERATION

A particle moves from point P to Q along a curvilinear path, as shown in Fig. 3.2

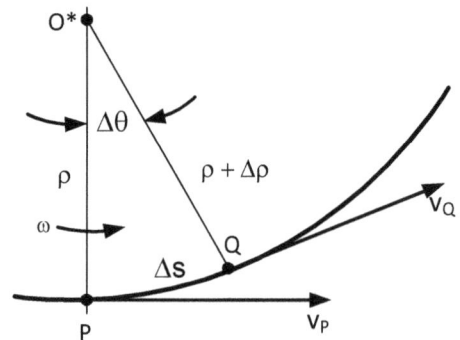

Fig. 3.2 A particle moves along the incremental arc Δs from point P to Q.

It is evident from previous developments that the acceleration is given by

$$\mathbf{a} = \lim_{t \to 0}\frac{\mathbf{\Delta v}}{\Delta t} \tag{3.25}$$

To determine the normal and tangential components of the acceleration, let's reference Fig. 3.2 and translate the vector v_Q to join vector v_P at a common point O_1, as shown in Fig. 3.3. On the vector triangle in Fig. 3.3, we measure the length of vector v_P along vector v_Q and establish point C. The vertices of the larger triangle are labeled O_1, A and B. We construct the line AC, which is perpendicular to the line O_1–B, to form the smaller triangle ACB.
The length of the line A-B is equal to Δv.

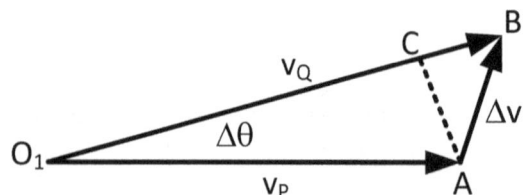

Fig. 3.3 The velocity triangle that was constructed using the results from Fig. 3.2.

From the geometry of the smaller triangle, we note that the vector AB is given by:

$$AB = \Delta v = AC + CB \tag{3.26}$$

Then it is evident that the acceleration vector **a** is given by:

$$\mathbf{a} = \lim_{t \to 0} \frac{\Delta v}{\Delta t} = \lim_{t \to 0} \frac{AC + CB}{\Delta t} = \lim_{t \to 0} \frac{AC}{\Delta t} + \lim_{t \to 0} \frac{CB}{\Delta t} \tag{3.27}$$

The term $\lim_{t \to 0} = \frac{AC}{\Delta t}$ represents the acceleration due to the change in the direction of the velocity of the particle, as it moves along the curvilinear path. The term $\lim_{t \to 0} = \frac{CB}{\Delta t}$ represents the acceleration due to the change in magnitude of the velocity, as the particle moves along the curvilinear path.

Let's explore the small triangle in Fig. 3.3 in more detail as Δt and $\Delta \theta$ go toward zero. As $\Delta \theta \to 0$, the angle $O_1AC \to 90^O$ and in the limit the line AC is perpendicular to both line O_1A and O_1C. Hence AC = $v\Delta\theta$ and we write:

$$a_n = \lim_{t \to 0} \frac{AC}{\Delta t} = v \lim_{t \to 0} \frac{\Delta\theta}{\Delta t} = v \frac{d\theta}{dt} = v\omega \tag{3.28}$$

where $\omega = d\theta/dt$ is the angular velocity of the radius of curvature ρ, shown in Fig. 3.2.

The term a_n is the component of the acceleration in the direction normal to the curvilinear path of the particle at point P. Reference to Fig. 3.2 indicates that the velocity v_t at point P is given by the radius ρ times the angular velocity ω. Hence, we may write:

$$v_t = \rho\omega \tag{3.29}$$

Recall $\alpha = d\omega/dt$ and then it is evident that:

$$a_t = \rho\alpha \tag{3.29a}$$

Then letting v in Eq. (3.28) be equal to v_t and substituting into (3.29) we find:

$$a_n = v_t \omega = \rho\omega^2 = v_t^2/\rho \tag{3.30}$$

Again reference Fig. 3.3 and note as Δt and $\Delta\theta \to 0$, the line CB represents the change in the magnitude of the velocity. Accordingly we may write:

$$a_t = \lim_{t \to 0} \frac{CB}{\Delta t} = \lim_{t \to 0} \frac{v_Q - v_P}{\Delta t} = \lim_{t \to 0} \frac{\Delta v}{\Delta t} = \frac{dv}{dt} \tag{3.31}$$

This is the component of the acceleration that is tangential to the curvilinear path at point P.

Because $v_t = ds/dt$, the tangential component of the acceleration may be expressed as:

$$a_t = d^2s/dt^2 \quad \text{or} \quad a_t = \rho\alpha \tag{3.32}$$

The magnitude of the acceleration vector is given by:

$$a = [a_t^2 + a_n^2]^{1/2} \tag{3.33}$$

and its direction is:

$$\mathbf{u_a} = \mathbf{a}/a \tag{3.34}$$

EXAMPLE 3.11

A particle is moving clockwise along a circular path that has a radius of 2 m. The angular velocity ω of the particle relative to the center of the circle is given by:

$$\omega = kt^2 \tag{a}$$

where k is a constant and ω is expressed in terms of radians/s.

If the tangential velocity of a particle is $v = 64$ m/s when $t = 2$ s, determine the tangential velocity and tangential and normal acceleration when $t = \frac{1}{2}$ s.

Solution

The tangential velocity is given by Eq. (3.29a) as:

$$v_t = r\omega = 2kt^2 \tag{b}$$

At $t = 2$ s, $v = 64$ m/s, which yields a solution for k as:

$$k = \frac{v_t}{2t^2} = \frac{64}{2(2)^2} = 8\,s^{-3} \tag{c}$$

From Eqs. (b) and (c) we write:

$$v_t = 16\,t^2 \tag{d}$$

Then at $t = \frac{1}{2}$ s we find:

$$v_t = 16(1/2)^2 = 4 \text{ m/s} \tag{e}$$

The tangential acceleration is given by:

$$a_t = \frac{dv_t}{dt} = \frac{d}{dt}(16t^2) = 32t \tag{f}$$

Then at $t = \frac{1}{2}$ s we find:

$$a_t = 16 \text{ m/s}^2 \tag{g}$$

$$a_n = v_t^2/r = (4)^2/2 = 8 \text{ m/s}^2 \tag{h}$$

EXAMPLE 3.12

A particle P is moving down a parabolic incline with a tangential velocity $v_t = 8$ ft/s and an tangential acceleration of 1.5 ft/s^2, as shown in Fig. E3.12. The parabola follows the equation $y = x^2/25$. Determine the direction of the velocity and the magnitude and direction of the total acceleration of the particle at the point P.

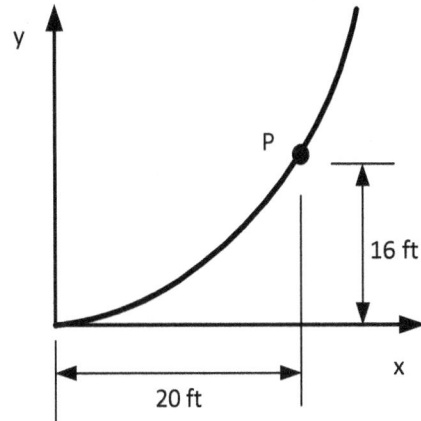

Fig. E3.12 A particle P moving down a parabolic shaped incline at the position indicate in the sketch.

Solution:

The particle's velocity is tangent to the parabolic incline; hence,

$$dy/dx = 2x/25 = x/12.5 \qquad \text{at } x = 20 \text{ ft} \qquad dy/dx = 20/12.5 = 1.6 \text{ ft/s} \qquad \text{(a)}$$

The term dy/dx represents the slope of the parabolic incline; hence, $\tan \theta = \tan (dy/dx)$, which is expressed as the $\tan^{-1}$ of this slope.

$$\theta = \tan^{-1} 1.6 = 58.0° \qquad \text{(b)}$$

The acceleration has tangential and normal components. The tangential acceleration was given in the problem statement as $a_t = 1.5$ ft/s^2. The normal acceleration is given by:

$$a_n = v_t^2/\rho \qquad \text{(c)}$$

where ρ is the radius of curvature of the parabola at the particle's location, which is given by:

$$\rho = \frac{\left[1 + \left(\dfrac{dy}{dx}\right)^2\right]^{3/2}}{\left|\dfrac{d^2y}{dx^2}\right|} \qquad \text{(d)}$$

Substituting numerical values into Eq. (d) gives:

$$\rho = \frac{\left[1+(1.6)^2\right]^{3/2}}{0.08} = 83.96 \text{ ft} \qquad (e)$$

Then from $a_n = v_t^2/\rho$:

$$a_n = (8)^2/83.96 = 0.7623 \text{ ft/s}^2 \qquad (f)$$

The magnitude of the acceleration is given by:

$$a = [(1.5)^2 + (0.7623)^2]^{1/2} = 1.683 \text{ ft/s}^2 \qquad (g)$$

The direction of the acceleration vector is:

$$\theta = \tan^{-1} a_n/a_t = \tan^{-1} 0.7623/1.5 = 26.94^\circ \qquad (h)$$

where α is defined in Fig. E3.12a:

Fig E3.12a The direction of the acceleration vector.

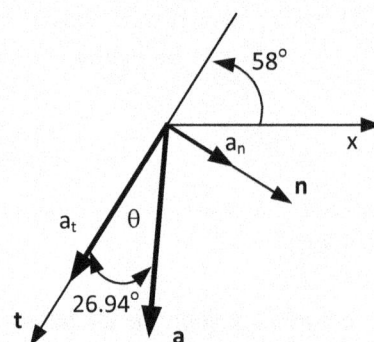

3.4 RADIAL AND TRANSVERSE COMPONENTS OF ACCELERATION

Consider a particle P in motion along a curvilinear path, as shown in Fig. 3.4, where we have drawn unit vectors $\mathbf{u}_r$ and $\mathbf{u}_T$ in the radial and transverse directions, respectively. As the particle moves along the curvilinear path by a small amount, the angle θ increases by $\Delta\theta$. With this incremental motion, the unit vectors $\mathbf{u}_r$ and $\mathbf{u}_T$ rotate, as shown in Fig. 3.5. This rotation of the unit vectors is important, because they must be considered in any time differentiation of the position vector to obtain the velocity and acceleration of the particle[2].

Fig. 3.4 Unit vectors $\mathbf{u}_r$ and $\mathbf{u}_T$ identify the radial and transverse directions, as the particle moves along the curvilinear path.

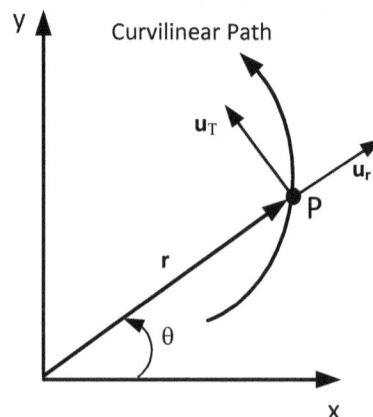

---

[2] The unit vectors $\mathbf{u}_r$ and $\mathbf{u}_T$ differ from the more common unit vectors $\mathbf{i}$, $\mathbf{j}$ and $\mathbf{k}$, which are fixed to the origin of a rectangular coordinate system. Instead the unit vectors $\mathbf{u}_r$ and $\mathbf{u}_T$ travel along a curvilinear path and while their magnitude remain fixed at unity, they rotate as they move with the particle P. This rotation causes their derivatives to differ from zero.

Let's begin by writing the relation for the position vector as:

$$\mathbf{r} = r\,\mathbf{u_r} \qquad (3.35)$$

The velocity is:

$$\mathbf{v} = d\mathbf{r}/dt = r\,d\mathbf{u_r}/dt + \mathbf{u_r}\,dr/dt \qquad (3.36)$$

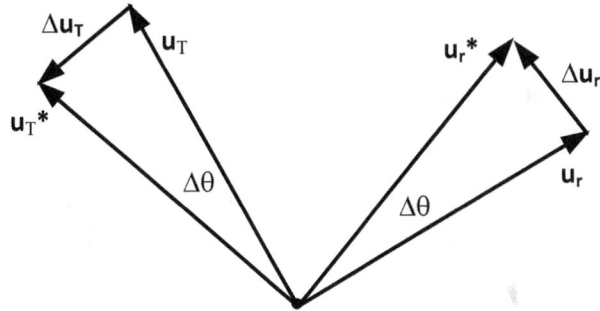

Fig. 3.5 As the particle moves along a curvilinear path the unit vectors $\mathbf{u_r}$ and $\mathbf{u_T}$ rotate.

Inspection of Fig. 3.5 shows that

$$\mathbf{u_r}^* = \mathbf{u_r} + \mathbf{\Delta u_r} \qquad (3.37)$$

Because $\mathbf{u_r}$ has a magnitude of one, it is evident that $\mathbf{\Delta u_r} = (1.0)\,\Delta\theta$ with its direction given by $\mathbf{u_T}$.

Considering the limit of these quantities as Δt goes to zero enables us to write:

$$\frac{d\mathbf{u_r}}{dt} = \lim_{t\to 0}\frac{\mathbf{\Delta u_r}}{\Delta t} = \lim_{t\to 0}\left(\frac{\Delta\theta}{\Delta t}\right)\mathbf{u_T} = \frac{d\theta}{dt}\mathbf{u_T} \qquad (3.38)$$

Combining Eqs. (3.35) and (3.36) gives:

$$\mathbf{v} = r\frac{d\theta}{dt}\mathbf{u_T} + \frac{dr}{dt}\mathbf{u_r} = v_T\mathbf{u_T} + v_r\mathbf{u_r} \qquad (3.39)$$

where $v_T = r\,d\theta/dt = r\omega$ and $v_r = dr/dt$.

The magnitude of the velocity is given by:

$$v = [(v_r)^2 + (v_T)^2]^{1/2} \qquad (3.40)$$

The velocity vector is tangent to the curvilinear path, upon which the particle is traveling. The components of the velocity in the radial and transverse directions are shown in Fig. 3.6.

Using Eq. (3.39) we write the acceleration of the particle as:

$$\mathbf{a} = d\mathbf{v}/dt = r\,d\theta/dt\,d\mathbf{u_T}/dt + [d/dt(r\,d\theta/dt)]\,\mathbf{u_T} + dr/dt\,d\mathbf{u_r}/dt + d^2r/dt^2\,\mathbf{u_r} \qquad (3.41)$$

We will show that this involved expression simplifies into two components with unit vectors providing directions in the r and T direction, as indicated in the equation below.

$$\mathbf{a} = a_r\, \mathbf{u}_r + a_T\, \mathbf{u}_T$$

Reference to Fig. 3.5 indicates that $\mathbf{u}_T^* = \mathbf{u}_T + \Delta\mathbf{u}_T$ due the rotation $\Delta\theta$ during the interval Δt. Taking the derivative of the unit vector $\mathbf{u}_T$ with respect to time gives:

$$\frac{d\mathbf{u}_T}{dt} = \lim_{t \to 0} \frac{\Delta\mathbf{u}_T}{\Delta t} \qquad\qquad (a)$$

For small angles $\Delta\mathbf{u}_T = (1.0)\Delta\theta$ and $\Delta\mathbf{u}_T$ acts in the negative $\mathbf{u}_R$ direction; hence, $\Delta\mathbf{u}_T = -\Delta\theta\mathbf{u}_R$. Substituting this result into Eq. (a) yields:

$$\frac{d\mathbf{u}_T}{dt} = -\lim_{t \to 0} \frac{\Delta\theta}{\Delta t}\mathbf{u}_r = -\frac{d\theta}{dt}\mathbf{u}_r \qquad\qquad (3.42)$$

The minus sign in Eq. (3.42) is because $\Delta\mathbf{u}_T$ acts in the negative $\mathbf{u}_r$ direction.

Fig. 3.6 Radial and transverse components of velocity at a point
along a curvilinear path of motion.

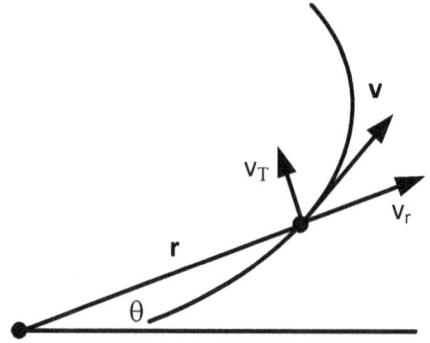

By substituting Eqs. (3.38) and (3.42) into Eq. (3.41), we determine:

$$\mathbf{a} = r\frac{d\theta}{dt}\left(-\frac{d\theta}{dt}\right)\mathbf{u}_r + \left(r\frac{d^2\theta}{dt^2} + \frac{d\theta}{dt}\frac{dr}{dt}\right)\mathbf{u}_T + \left(\frac{dr}{dt}\frac{d\theta}{dt}\right)\mathbf{u}_T + \frac{d^2r}{dt^2}\mathbf{u}_r \qquad (3.43)$$

We separate terms in Eq. (3.43) and write the expression for the acceleration $\mathbf{a}$ in terms of radial and transverse components as:

$$\mathbf{a} = a_r\, \mathbf{u}_r + a_T\, \mathbf{u}_T \qquad\qquad (3.44)$$

where the magnitude of the radial and transverse components are given by:

$$a_r = [d^2r/dt^2 - r(d\theta/dt)^2] \qquad \text{and} \qquad a_T = [r\, d^2\theta/dt^2 + 2(d\theta/dt)(dr/dt)] \qquad (3.45)$$

The magnitude of the acceleration vector is given by:

$$a = \{[d^2r/dt^2 - r(d\theta/dt)^2]^2 + [r\, d^2\theta/dt^2 + 2(d\theta/dt)(dr/dt)]^2\}^{1/2} \qquad (3.46)$$

The direction of the acceleration vector is evident from its radial and transverse components, as shown in Fig. 3.7.

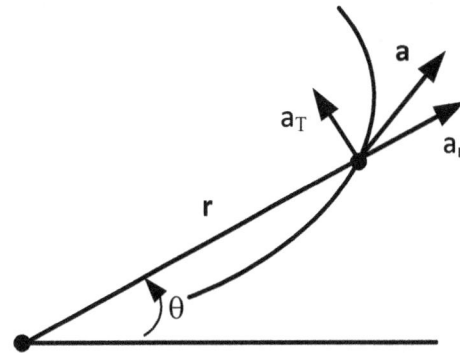

Fig. 3.7 Graphical representation of typical acceleration
vector and its radial and transverse components.

EXAMPLE 3.13

A linear bearing slides along a rod with a cylindrical cross section, as shown in Fig. E3.13. The rod, initially in alignment with the x axis, is rotating about the origin with its orientation angle θ given by: $\theta = 0.30\ t^2$, where θ is expressed in radians and t in seconds. The position of the linear bearing along the rod is given by $r = 1.8 - 1.5t^2$, where r is in ft. When the rod has rotated to $\theta = 15°$, determine the velocity and acceleration of the bearing.

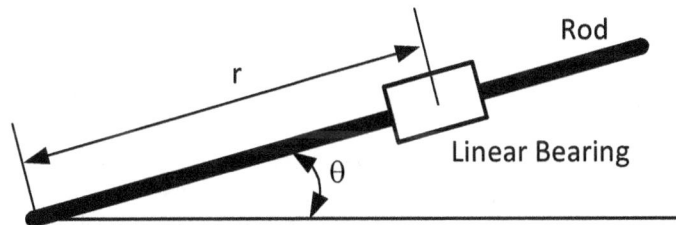

Fig. E3.13 A linear bearing moving along
a rotating rod.

Solution:

Let's determine the time when the rod is at $\theta = 15°$ by:

$$t^2 = \theta/0.30 = [(15/180)\ \pi]/0.30 = 0.8727\ s^2 \qquad \text{and}\ t = 0.9342\ s \qquad (a)$$

Next let's find the radial position, velocity and acceleration of the bearing by:

$$r = 1.8 - 1.5\ t^2 = 1.8 - 1.5(0.8727) = 0.4910\ \text{ft} \qquad (b)$$

$$v_r = dr/dt = -3.0\ t = -3.0(0.9342) = -2.803\ \text{ft/s} \qquad (c)$$

$$d^2r/dt^2 = -3.0\ \text{ft/s}^2 \qquad (d)$$

Let's find the transverse position, velocity and acceleration of the bearing B by:

$$\theta = 0.3\ t^2 = 0.3\ (0.8727) = 0.2618\ \text{radians} \qquad (e)$$

$$d\theta/dt = 0.6\ t = 0.6\ (0.9342) = 0.5605\ \text{radian/s} \qquad (f)$$

$$d^2\theta/dt^2 = 0.6 \text{ radian/s}^2 \tag{g}$$

The total velocity of the bearing B is given by:

$$v_r = dr/dt = -2.803 \text{ ft/s} \quad \text{and} \quad v_T = r\,(d\theta/dt) = 0.4910\,(0.5605) = 0.2752 \text{ ft/s} \tag{h}$$

$$v_B = [(v_r)^2 + (v_T)^2]^{1/2} = [(-2.803)^2 + (0.2752)^2]^{1/2} = 2.816 \text{ ft/s} \tag{i}$$

The direction of the velocity v_B is:

$$\theta = \tan^{-1} v_T/v_r = \tan^{-1}(0.2752/-2.803) = -5.607° \tag{j}$$

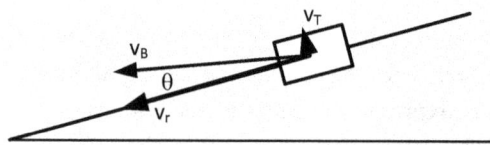

The total acceleration of the bearing B is:

$$a_r = d^2r/dt^2 - r(d\theta/dt)^2 = -3.0 - 0.4910(0.5605)^2 = -3.154 \text{ ft/s}^2 \tag{k}$$

$$a_T = r\,d^2\theta/dt^2 + 2\,(dr/dt)(d\theta/dt) = 0.4910(0.6) + 2(-2.803)(0.5605) = -2.848 \text{ ft/s}^2 \tag{l}$$

$$a_B = [(a_r)^2 + (a_T)^2]^{1/2} = [(-3.154)^2 + (-2.848)^2]^{1/2} = 4.250 \text{ ft/s} \tag{m}$$

The direction of the total acceleration vector is given by:

$$\theta = \tan^{-1} a_T/a_r = \tan^{-1}(-2.848/-3.154) = \tan^{-1} 0.9030 = 42.08° \tag{n}$$

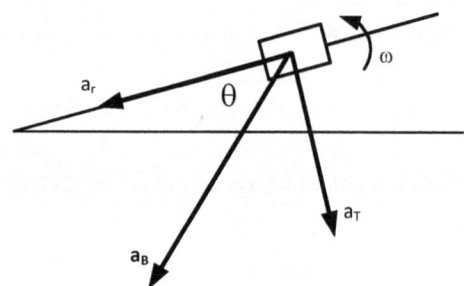

EXAMPLE 3.14

The rod shown in Fig. E3.14 houses a spring that maintains a pin in contact with a parabolic shaped cam. The relation specifying the cam's shape is:

$$r = 2d/(1 + \cos \theta) \tag{a}$$

The angle that the rod makes with the x axis is given by:

$$\theta = kt \tag{b}$$

Determine the equations describing the radial and transverse velocity and acceleration components of the pin when $\theta = 0°$ and $90°$.

Fig. E3.14 The mechanism for a pin to follow a parabolic cam path.

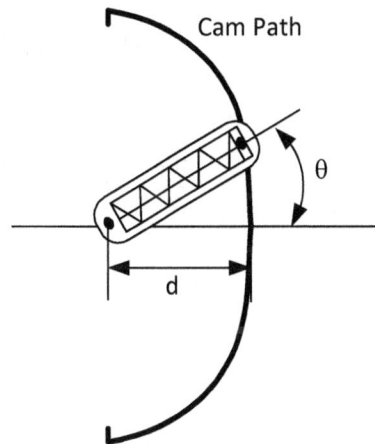

Cam Path

θ

d

Solution

Let's begin by determining the first and second derivatives of r and θ with respect to t.

$$\theta = kt, \qquad d\theta/dt = k \qquad \text{and} \qquad d^2\theta/dt^2 = 0 \tag{c}$$

$$r = 2d/(1 + \cos \theta)$$

$$\frac{dr}{dt} = 2d\left[\frac{(1+\cos\theta)(0) - (1)(-\sin\theta)(d\theta/dt)}{(1+\cos\theta)^2}\right] = 2d\left[\frac{\sin\theta(d\theta/dt)}{(1+\cos\theta)^2}\right] \tag{d}$$

$$\frac{d^2r}{dt^2} = 2d\left[\frac{(1+\cos\theta)^2[\sin\theta(d^2\theta/dt^2) + \cos\theta(d\theta/dt)^2] - \sin\theta(d\theta/dt)(2)(1+\cos\theta)(-\sin\theta)(d\theta/dt)}{(1+\cos\theta)^4}\right]$$

that reduces to:

$$\frac{d^2r}{dt^2} = 2d \left[\frac{(1+\cos\theta)^2[\sin\theta(d^2\theta/dt^2) + \cos\theta(d\theta/dt)^2] + 2\sin^2\theta(d\theta/dt)^2(1+\cos\theta)}{(1+\cos\theta)^4} \right] \tag{e}$$

When $\theta = 0$ then

$$r = d \tag{f}$$

$$\frac{dr}{dt}_{\theta=0} = 2d \left[\frac{\sin\theta(d\theta/dt)}{(1+\cos\theta)^2} \right] = 2d \left[\frac{(0)k}{(1+1)^2} \right] = 0 \tag{g}$$

$$\frac{d^2r}{dt^2}_{\theta=0} = 2d \left[\frac{(1+1)^2[(0)(0)+(1)k^2]+2(0)k^2(1+1)}{(1+1)^4} \right] = \frac{dk^2}{2} \tag{h}$$

when $\theta = 90°$

$$r = 2d \tag{i}$$

$$\frac{dr}{dt}_{\theta=90} = 2d \left[\frac{\sin\theta(d\theta/dt)}{(1+\cos\theta)^2} \right] = 2d \left[\frac{(1)k}{(1)^2} \right] = 2dk \tag{j}$$

$$\frac{d^2r}{dt^2}_{\theta=90} = 2d \left[\frac{(1)^2[(1)(0)+(0)k^2]+2(1)^2k^2(1)}{(1)^4} \right] = 4dk^2 \tag{k}$$

When $\theta = 0^0$ the velocity components are:

$$v_r = dr/dt = 0 \tag{l}$$

$$v_T = r \, d\theta/dt = dk \tag{m}$$

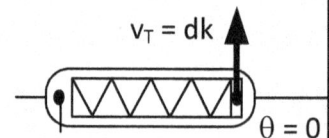

When $\theta = 90^0$ the velocity components are:

$$v_r = dr/dt = 2dk \qquad \text{and} \qquad v_T = r \, d\theta/dt = 2dk \tag{n}$$

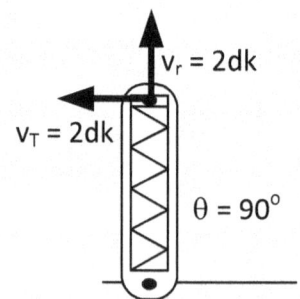

When $\theta = 0^0$ the acceleration components are:

$$a_r = d^2r/dt^2 - r(d\theta/dt)^2 = dk^2/2 - dk^2 = -dk^2/2 \tag{o}$$

$$a_T = r \, d^2\theta/dt^2 + 2(dr/dt)(d\theta/dt) = d(0) + 2(0)k = 0 \tag{p}$$

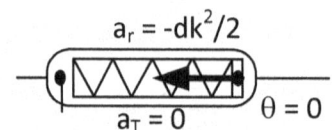

When $\theta = 90^0$ the acceleration components are:

$$a_r = d^2r/dt^2 - r\,(d\theta/dt)^2 = 4dk^2 - 2dk^2 = 2dk^2 \qquad (q)$$

$$a_T = r\,d^2\theta/dt^2 + 2\,(dr/dt)\,(d\theta/dt) = 2d(0) + 2(2dk)k = 4dk^2 \qquad (r)$$

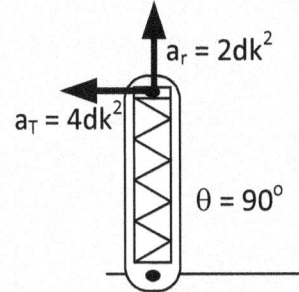

$a_r = 2dk^2$

$a_T = 4dk^2$

$\theta = 90°$

3.5 ANGULAR ACCELERATION

Consider a particle moving with a velocity v along a curvilinear path, as shown in Fig. 3.8. The particle is located by a radial line ρ from its instantaneous center and the radial line is rotating with an angular velocity ω. Recall with a curved path, the curvature may change as the particle moves along the path. Because of the change in curvature, the instantaneous center changes position and the radius of curvature changes length. Do not confuse this curved path with a circle, where the curvature is constant, the center of curvature is fixed and the radius in a constant as the particle moves about the circle.

The velocity vector is tangent to the curved path at point P, as the particle moves along the curved path. The angular acceleration α of this line is the time rate of change of the angular velocity ω, which is given by:

$$\alpha = \lim_{t \to 0} \frac{\Delta\omega}{\Delta t} = \frac{d\omega}{dt} \qquad (3.47)$$

and

$$\alpha = \frac{d\omega}{dt} = \frac{d^2\theta}{dt^2} \qquad (3.48)$$

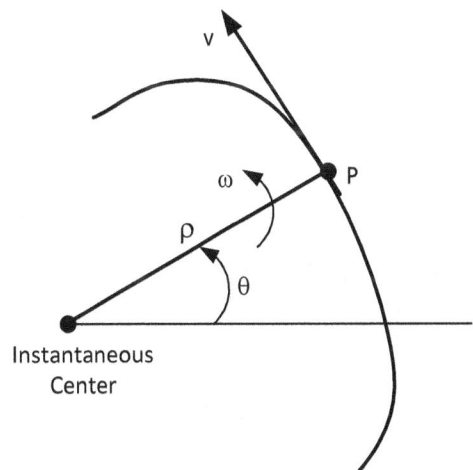

Fig. 3.8 A particle a point P is located by position parameters ρ and θ.

Instantaneous Center

3.6. PARTICLE MOTION ALONG A CIRCULAR PATH

In many engineering application particles move along circular paths. Examples include points on wheels, shafts, gears, propellers, turbine discs, etc. It is important to recognize this important type of motion and to note the relations among the quantities of interest, which include the angular velocity ω and acceleration α as well as the position θ. When a particle moves along a circular path, the center of curvature is fixed at the center of the circle and $\rho = r = $ constant. In this case, the transverse components of velocity acceleration coincide with the tangential components of velocity and acceleration. However, the direction of the radial component of acceleration is opposite to the direction of the normal component of acceleration if the polar origin is at the center of the circle.

If the angular acceleration is a constant with respect to time, we may write:

$$\omega = \omega_0 + \alpha t \qquad (3.49)$$

Integrating Eq. (3.49) and setting the constant of integration equal to zero yields:

$$\theta = \omega_0 t + \tfrac{1}{2} \alpha t^2 \qquad (3.50)$$

Square Eq. (3.49) to obtain:

$$\omega^2 = \omega_0^2 + 2\omega_0 \alpha t + \alpha^2 t^2 \qquad (a)$$

Solving Eq. (3.50) for t^2 gives:

$$t^2 = (\theta - \omega_0 t)(2/\alpha) \qquad (b)$$

Substitute Eq. (b) into Eq. (a) and reducing yields:

$$\omega^2 = \omega_0 + 2\alpha\theta \qquad (3.51)$$

We may also develop a relationship between linear and angular acceleration for particle moving along a circular path. To illustrate these relations, consider the particle moving along a circular path, as shown in Fig. 3.9.

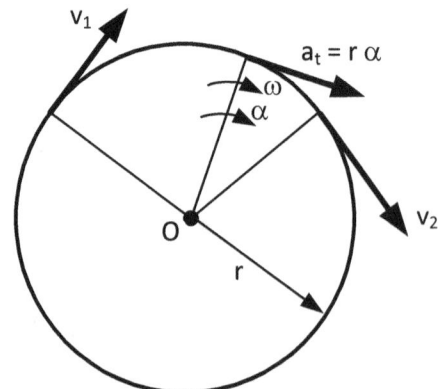

Fig. 3.9 A particle moving along a circular path with a radius r.
In this case the radius of curvature ρ is a constant equal to r.

Let's recall that $a_t = dv/dt$ and $v = r\omega$; hence:

$$a_t = \frac{d}{dt}(r\omega) = r\frac{d\omega}{dt} \tag{3.52}$$

Recall that $\alpha = d\omega/dt$, then:

$$a_t = r\,\alpha \tag{3.53}$$

Equation (3.53) is also valid if the path is not circular, provided that ρ is the radius of curvature of the curved path at the particle's location and α is the angular acceleration of the particle with respect to the center of curvature.

The normal component of acceleration $a_n = r\omega^2$ is independent from the tangential component of the acceleration. Instead a_n depends on ω and not on α at the instant under consideration.

EXAMPLE 3.15

A particle moves on a circular path with $\theta = 3t + 2$. Determine the angular velocity ω and the angular acceleration α as a function of time. Also determine the angular velocity ω and the angular acceleration α at $t = 3$ s.

Solution

We determine the relation for the angular velocity ω by differentiating θ with respect to time t.

$$\omega = \frac{d\theta}{dt} = \frac{d}{dt}(3t + 2) = 3 \tag{a}$$

We determine the relation for the angular acceleration α by differentiating ω with respect to time t.

$$\alpha = \frac{d\omega}{dt} = \frac{d}{dt}(3) = 0 \text{ radians}/s^2 \tag{b}$$

These results indicate that the particle is moving about the circular path with a constant angular velocity ω of 3 rad/s and with an angular acceleration of zero.

It is evident for anytime t that $\alpha = 0$ rad/s^2 and $\omega = 3$ rad/s.

EXAMPLE 3.16

The flywheel, shown in Fig. E3.16, is 6 ft in diameter and is rotating initially with an angular velocity $\omega = 120$ RPM. During a 6 s period, it undergoes a uniform change in angular velocity until it slows to 60 RPM. Determine the tangential component of acceleration of a point on the rim of the flywheel during this 6 s period. Find the total acceleration of a point on the rim of the flywheel at the end of this 6 s period.

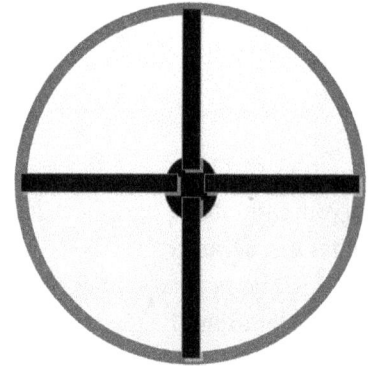

Fig. E3.16 The flywheel with a diameter of 6 ft is rotating with an initial angular velocity ω = 120 RPM. Its angular velocity decreases at a uniform rate to 60 RPM over a 6 s period

---

Solution

Before beginning the computations required, let's convert RPM to radians per second.

$$\frac{60\text{Rev}}{\text{Min}} \times \frac{2\pi}{\text{Rev}} \times \frac{\text{Min}}{60\text{s}} = 2\pi \text{ rad} / \text{s}$$

(a)

Then it is clear that   120 RPM = 4π rad/s.

Because the angular velocity decreases at a uniform rate, we know that the difference in the tangential velocities divided by Δt is equal to the tangential acceleration.

$$a_t = r\frac{\omega_1 - \omega_0}{\Delta t} = \frac{6}{2}\left(\frac{2\pi - 4\pi}{6}\right) = -\pi \text{ ft} / s^2 = -3.142 \text{ ft/s}^2$$

(b)

The normal component of the acceleration at the end of the 6 s period is determined from Eq. (3.30) as:

$$a_n = r\,\omega^2 = 3\,(2\pi)^2 = 12\,\pi^2 \text{ ft/s}^2 = 118.4 \text{ ft/s}^2$$

(c)

The total acceleration is determined from Eq. (3.30) as:

$$a = \sqrt{(a_t)^2 + (a_n)^2} = \sqrt{(-\pi)^2 + (12\pi^2)^2} = 118.5 \text{ ft./s}^2$$

(d)

The orientation of the acceleration vector of a particle on the rim of the flywheel at the end of the 6 second period is:

$$\theta = \tan^{-1}\frac{a_t}{a_n} = \tan^{-1}\left(\frac{-\pi}{12\pi^2}\right) = \tan^{-1}\left(\frac{-1}{12\pi}\right) = \tan^{-1}(-0.02653) = -1.520^O$$

(e)

Note that the acceleration vector only deviates slightly from the normal direction, as shown in Fig. E3.16a.

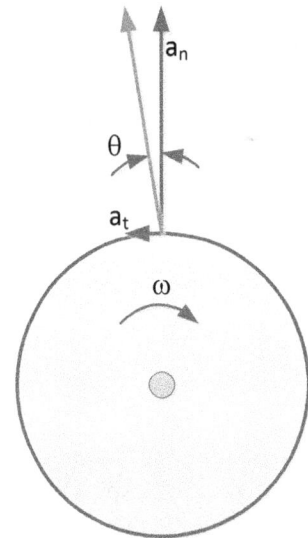

Fig. E3.16a Tangential, normal and total acceleration vectors at a point on
the rim of the flywheel as it reaches 60 RPM.

EXAMPLE 3.17

For the mechanism, shown in Fig. E3.17, determine the equations for the displacement x, the velocity v
and acceleration a of the linear bearing. The bearing is driven by a crank with a radius R that rotates
clockwise about a circle at a constant angular velocity ω.

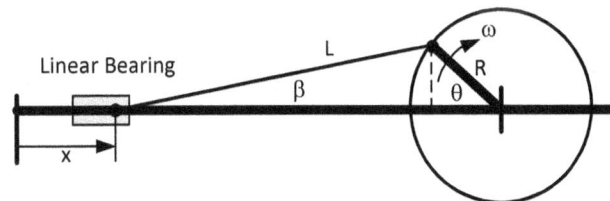

Fig. E3.17 Reciprocating linear bearing driven
by a rotating crank.

Solution:

The linear bearing oscillates along the shaft defined by the x axis, as the crank rotates with a constant
angular velocity ω. The linear bearing is at the far left hand position where x = 0 when β and θ both are
equal to zero. When the crank rotates clockwise, the linear bearing slides to the right and the geometry of
the mechanism enables us to express x as:

$$x = L + R - (L \cos \beta + R \cos \theta) \tag{a}$$

It is evident from the sketch in Fig. E3.17 that the height of the dotted line can be expressed as:

$$R \sin \theta = L \sin \beta \qquad \text{or} \qquad \sin \beta = (R/L) \sin \theta \tag{b}$$

Squaring both sides of the second relation in Eq. (b) gives:

$$\sin^2 \beta = (R/L)^2 \sin^2 \theta \qquad \text{or} \qquad (1 - \cos^2 \beta) = (R/L)^2 \sin^2 \theta \tag{c}$$

Then

$$\cos \beta = [1 - (R/L)^2 \sin^2 \theta]^{1/2} \tag{d}$$

To simplify the differentiation of the expression for x, expand the right side of Eq. (d) in a power series[3]:

$$\cos \beta = [1 - (R/L)^2 \sin^2 \theta]^{1/2} = 1 - \tfrac{1}{2} (R/L)^2 \sin^2 \theta \tag{e}$$

Substituting Eq. (e) into Eq. (a) yields:

$$x = L + R - L[1 - \tfrac{1}{2} (R/L)^2 \sin^2 \theta] - R \cos \theta \tag{f}$$

We simplify Eq. (f) and write:

$$x = R (1 - \cos \theta) + (R^2/2L) \sin^2 \theta] \tag{g}$$

We derive the expression for the velocity v of the linear bearing by differentiating x with respect to t and noting that $dR/dt = 0$, because the crank radius is fixed:

$$v = dx/dt = R \sin \theta (d\theta/dt) + (R^2/2L)[2\sin \theta \cos \theta (d\theta/dt)] \tag{h}$$

Set $d\theta/dt = \omega$, and use the trigometric identity that $\sin 2\theta = 2 \sin \theta \cos \theta$ to reduce Eq. (h) to:

$$v = R\omega [\sin \theta + (R/2L) \sin 2\theta] \tag{i}$$

Differentiating velocity with respect to t yields the acceleration of the linear bearing along the shaft as:

$$a = dv/dt = R\omega [\cos \theta (d\theta/dt) + (R/L) \cos 2\theta (d\theta/dt)] \tag{j}$$

$$a = R\omega^2 [\cos \theta + (R/L) \cos 2\theta] \tag{k}$$

The results for both the velocity and the acceleration of the linear bearing are approximations, because we only considered the first term in the power series expansion used in Eq. (e).

3.7 SUMMARY

The position of a particle moving along a curvilinear path is space is given by a position vector **r**:

$$\mathbf{r} = x\mathbf{i} + y\mathbf{j} + z\mathbf{k} \tag{3.1}$$

The velocity vector **v** is determined by:

$$\mathbf{v} = dx/dt \, \mathbf{i} + dy/dt \, \mathbf{j} + dz/dt \, \mathbf{k} \tag{3.2}$$

---

[3] Power series expansions are described in Calculus textbooks. For example see "Advanced Calculus for Engineers" by F. B. Hildebrand, Prentice Hall, page 121, 1949.

where $\qquad v_x = dx/dt \qquad\qquad v_y = dy/dt \qquad\qquad v_z = dz/dt$

or $\qquad\qquad\qquad\qquad \mathbf{v} = v_x\,\mathbf{i} + v_y\,\mathbf{j} + v_z\,\mathbf{k}$ $\qquad\qquad$ (3.3)

The magnitude of the velocity vector is:

$$v = [v_x^2 + v_y^2 + v_z^2]^{1/2} \qquad\qquad (3.4)$$

The direction of the velocity vector is expressed in terms of the unit tangent vector $\mathbf{u_v}$ as:

$$\mathbf{u_v} = \mathbf{v}/v \qquad\qquad (3.5)$$

The acceleration of the particle is:

$$\mathbf{a} = d^2x/dt^2\,\mathbf{i} + d^2y/dt^2\,\mathbf{j} + d^2z/dt^2\,\mathbf{k} \qquad\qquad (3.6)$$

where $\qquad a_x = d^2x/dt^2 \qquad\qquad a_y = d^2y/dt^2 \qquad\qquad a_z = d^2z/dt^2$

or $\qquad\qquad\qquad\qquad \mathbf{a} = a_x\,\mathbf{i} + a_y\,\mathbf{j} + a_z\,\mathbf{k}$ $\qquad\qquad$ (3.7)

The magnitude of the acceleration vector is:

$$a = [a_x^2 + a_y^2 + a_z^2]^{1/2} \qquad\qquad (3.8)$$

The direction of the acceleration vector is expressed in terms of a unit vector $\mathbf{u_a}$ as:

$$\mathbf{u_a} = \mathbf{a}/a \qquad\qquad (3.9)$$

Planar motion is a special case of three dimensional curvilinear motion; hence,

$$\mathbf{r} = x\mathbf{i} + y\mathbf{j} \qquad\qquad (3.10)$$

$$\mathbf{v} = dx/dt\,\mathbf{i} + dy/dt\,\mathbf{j} \qquad\qquad (3.11)$$

$$\mathbf{v} = v_x\,\mathbf{i} + v_y\,\mathbf{j} \qquad\qquad (3.12)$$

$$v = [v_x^2 + v_y^2]^{1/2} \qquad\qquad (3.13)$$

$$\theta = \tan^{-1} v_y/v_x \qquad\qquad (3.14)$$

$$\mathbf{a} = d^2x/dt^2\,\mathbf{i} + d^2y/dt^2\,\mathbf{j} \qquad\qquad (3.15)$$

$$\mathbf{a} = a_x\,\mathbf{i} + a_y\,\mathbf{j} \qquad\qquad (3.16)$$

$$a = [a_x^2 + a_y^2]^{1/2} \qquad\qquad (3.17)$$

$$\theta = \tan^{-1} a_y/a_x \qquad\qquad (3.18)$$

Projectile motion is a special case of planar curvilinear motion; hence,

The exit velocities of the projectile are:

$$v_x)_0 = v_{ex} \cos \theta \qquad (3.19a)$$

and

$$v_y)_0 = v_{ex} \sin \theta \qquad (3.19b)$$

After launch, the velocity of the projectile in the y direction is:

$$v_y = v_y)_0 - gt \qquad (3.20)$$

The relation for the height of the projectile, if the gun tube is initially located on the ground is:

$$y = v_y)_0 \, t - gt^2/2 \qquad (3.22)$$

The time required for the projectile to impact the ground plane is:

$$t_{Impact} = \frac{2v_y)_0}{g} \qquad (3.23)$$

The range x is:

$$x = v_x)_0 \, t_{Impact} \qquad (3.24)$$

For normal and tangential components of acceleration, we find:

The tangential velocity v_t is:

$$v_t = \rho \omega \qquad (3.29)$$

The normal and tangential acceleration components are:

$$a_n = v_t \, \omega = \rho \omega^2 = v_t^2/\rho \qquad (3.30)$$

$$a_t = d^2s/dt^2 \qquad (3.32)$$

The acceleration magnitude is:

$$a = [a_t^2 + a_n^2]^{1/2} \qquad (3.33)$$

The direction of the acceleration vector is:

$$\theta = \tan^{-1} \frac{a_y}{a_x} \qquad (3.34)$$

For radial and transverse components of velocity and acceleration the position vector is:

$$\mathbf{r} = r\,\mathbf{u_r} \tag{3.35}$$

The velocity is:

$$\mathbf{v} = d\mathbf{r}/dt = r\,d\mathbf{u_r}/dt + dr/dt\,\mathbf{u_r} \tag{3.36}$$

The derivative of the unit vector $\mathbf{u_r}$ is:

$$\frac{d\mathbf{u_r}}{dt} = \lim_{t \to 0} \frac{\Delta\mathbf{u_r}}{\Delta t} = \lim_{t \to 0} \left(\frac{\Delta\theta}{\Delta t}\right)\mathbf{u_T} = \frac{d\theta}{dt}\mathbf{u_T} \tag{3.38}$$

The total velocity vector is:

$$\mathbf{v} = r\frac{d\theta}{dt}\mathbf{u_T} + \frac{dr}{dt}\mathbf{u_r} = v_T\mathbf{u_T} + v_r\mathbf{u_r} \tag{3.39}$$

where $v_T = r\,d\theta/dt = r\omega$ and $v_r = dr/dt$ and the magnitude of the velocity is given by:

$$v = [(v_r)^2 + (v_T)^2]^{1/2} \tag{3.40}$$

The velocity vector is tangent to the curvilinear path upon which the particle is traveling.

The total acceleration vector is:

$$\mathbf{a} = d\mathbf{v}/dt = r\,d\theta/dt\,d\mathbf{u_T}/dt + d/dt(r\,d\theta/dt)\,\mathbf{u_T} + dr/dt\,d\mathbf{u_r}/dt + d^2r/dt^2\,\mathbf{u_r} \tag{3.41}$$

We showed the derivative of the unit vector $\mathbf{u_T}$ is:

$$\frac{d\mathbf{u_T}}{dt} = \lim_{t \to 0} \frac{\Delta\mathbf{u_T}}{\Delta t} = -\lim_{t \to 0} \frac{\Delta\theta}{\Delta t}\mathbf{u_r} = -\frac{d\theta}{dt}\mathbf{u_r} \tag{3.42}$$

The total acceleration vector is:

$$\mathbf{a} = r\frac{d\theta}{dt}\left(-\frac{d\theta}{dt}\right)\mathbf{u_r} + \left(r\frac{d^2\theta}{dt^2} + \frac{d\theta}{dt}\frac{dr}{dt}\right)\mathbf{u_T} + \left(\frac{dr}{dt}\frac{d\theta}{dt}\right)\mathbf{u_T} + \frac{d^2r}{dt^2}\mathbf{u_r} \tag{3.43}$$

$$\mathbf{a} = a_r\,\mathbf{u_r} + a_T\,\mathbf{u_T} \tag{3.44}$$

$$a_r = [d^2r/dt^2 - r(d\theta/dt)^2] \quad\text{and}\quad a_T = [r\,d^2\theta/dt^2 + 2(d\theta/dt)(dr/dt)] \tag{3.45}$$

$$a = \{[d^2r/dt^2 - r(d\theta/dt)^2]^2 + [r\,d^2\theta/dt^2 + 2(d\theta/dt)(dr/dt)]^2\}^{1/2} \tag{3.46}$$

And the angular acceleration is:

$$\alpha = \frac{d\omega}{dt} = \frac{d^2\theta}{dt^2} \tag{3.47}$$

For particle motion along a circular path:

$$\omega = \omega_0 + \alpha t \tag{3.49}$$

$$\theta = \omega_0 t + \tfrac{1}{2} \alpha t^2 \tag{3.50}$$

$$\omega^2 = \omega_0 + 2\alpha\theta \tag{3.51}$$

$$a_t = \frac{d}{dt}(r\omega) = r\frac{d\omega}{dt} \tag{3.52}$$

$$a_t = r\,\alpha \tag{3.53}$$

CHAPTER 4

KINEMATICS OF RIGID BODIES IN PLANAR MOTION

INTRODUCTION:

In Chapter 2 and 3, we dealt with one, two and three dimensional motion of particles. In this chapter we will consider rigid bodies instead of particles and restrict the body's motion to the x-y plane. We will develop the kinematic equations for rigid bodies with x-y-z (rectangular), n-t (normal tangential), and r-T (θ) (radial transverse) coordinate systems. We will also introduce the very important concept of relative motion.

The study of planar motion of rigid bodies is often divided into four topics including:

1. Rectilinear translation or straight line motion from one location to another without rotation of the rigid body.
2. Curvilinear translation or motion along a curved planar path from one location to another without rigid body rotation.
3. Rotation of the rigid body about a fixed axis.
4. General planar motion where a rigid body moves along a curvilinear path and rotates in the process.

CONCEPT PROBLEM FOR CHAPTER 4

Consider a mass of 0.8 kg attached to a strong cable with a length of 500 mm that is rotating in the vertical plane, as shown below. Prepare free body diagrams of the weight in the 12:00, 6:00, 3:00 and 9:00 o'clock positions. Determine the angular velocity of the mass, if the tension force T in the cable is zero, when it is in the 12 o'clock position. Determine the angular velocity of the mass, if the tension force T in the cable is twice the weight of the mass, when it is in the 12 o'clock position. Determine the angular acceleration of the mass when the body is in the 3:00 and 9:00 positions.

The mass and cable is rotating with an angular velocity ω.

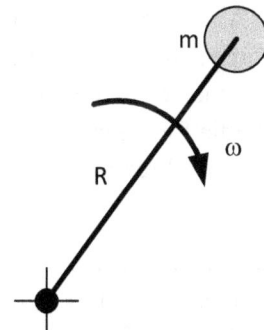

Let's begin by drawing a free body diagram showing the forces acting on the mass, when it is the 12 o'clock position. In this example, an inertia force[1] is added to the free body diagram. The inertia force is due to the acceleration a_n and it is directed radially outward and equal to ma_n. The advantage of adding an inertia force ma_n to the free body diagram is that it produces equilibrium of the body. Because the body is in equilibrium, we can use the equilibrium equations from Statics.

The free body diagram of the mass in the 12 o'clock position. An inertia force has been added to produce a body in equilibrium. The inertia force ma_n is shown with the dotted arrow in the positive n direction. We have not included an inertia force ma_t in the transverse direction, because no forces act on the body in this direction and at this instant $a_t = 0$ and the angular acceleration $\alpha = 0$.

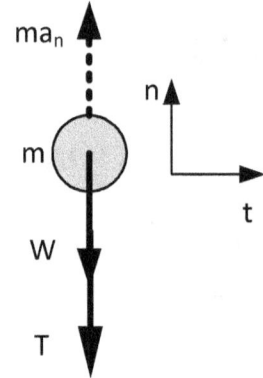

Summation of the forces in the n direction gives:

$$\Sigma F_n = ma_n - W - T = 0$$

$$ma_n = W \qquad \text{if } T = 0$$

$$mr\omega^2 = mg$$

$$\omega = (g/r)^{1/2}$$

$$\omega = (g/R)^{1/2} = (9.807/0.5)^{1/2} = 4.429 \text{ rad/s}$$

$$4.429 \text{ rad/s} \times \text{rev}/2\pi \text{ rad} \times 60 \text{ s/minute} = 42.29 \text{ rev/minute}$$

A free body diagram showing the forces acting on the mass when it is the 6 o'clock position is shown below. The inertia force ma_n has been added to the free body diagram to produce equilibrium of the body.

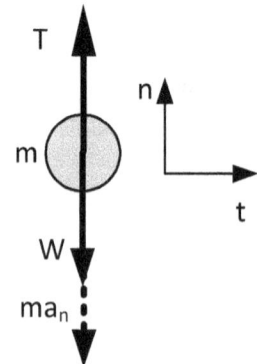

A free body diagram of the mass in the 6 o'clock position with the inertia force added to produce a body in equilibrium. The inertia force ma_n is shown in the negative n direction with the dotted arrow. We have not included an inertia force ma_t in the transverse direction, because no forces act on the body in this direction and at this instant $a_t = 0$ and the angular acceleration $\alpha = 0$.

----

[1] Inertia forces are sometimes called D'Alembert forces after D'Alembert a French mathematician who is credited with developing the concept in in the mid-1700s.

Summation of forces in the n direction gives:

$$\Sigma F_n = T - W - ma_n = 0$$

$$ma_n = W \qquad \text{if } T = 2W$$

$$mr\omega^2 = mg$$

$$\omega = (g/r)^{1/2}$$

$$\omega = (g/R)^{1/2} = (9.807/0.5)^{1/2} = 4.429 \text{ rad/s}$$

A free body diagram showing the forces acting on the mass when it is the 3 o'clock position is shown below. The inertia forces ma_n and ma_t have been added to the free body diagram to produce equilibrium of the body.

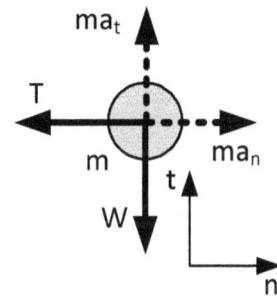

A free body diagram showing the forces acting on the mass, when it is the 3 o'clock position. The inertia force ma_n in the positive n direction is shown with the dotted arrow.

Summation of forces in the n and t directions gives:

$$\Sigma F_n = ma_n - T = 0$$

$$T = ma_n = mr\omega^2$$

$$\Sigma F_t = ma_t - W = 0$$

$$ma_t = mg$$

$$a_t = g = r\alpha$$

$$\alpha = g/r$$

A free body diagram showing the forces acting on the mass when it is the 9 o'clock position is shown below. The inertia forces ma_n and ma_t have been added to the free body diagram to produce equilibrium of the body.

A free body diagram showing the forces acting on the mass when it is the 9 o'clock position. The inertia force ma_n in the negative n direction is shown with the dotted arrow.

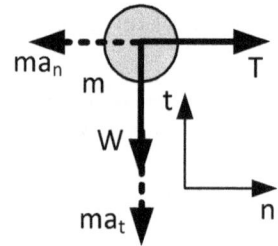

Summation of forces in the n and t directions gives:

$$\Sigma F_n = T - ma_n = 0$$

$$T = ma_n = mr\omega^2$$

$$\Sigma F_t = -ma_t - W = 0$$

$$ma_t = -mg$$

$$a_t = -g = r\alpha$$

$$\alpha = -g/r$$

We trust this simple problem enables you to understand that the normal acceleration of a rotating body is directed inward toward the center of rotation. The inertia force, added to the free body diagram, enables the use of the equilibrium equations, is in the opposite (outward) direction.

DISCUSSION

The development of centrifugal forces due to rotation of an object is fascinating. I recall as a youngster rotating a bucket of water in a vertical plane and not having a drop spill. I did not understand anything about gravitational forces or normal accelerations at the time, but I recognized forces developed when you rotated an object. Another example was when we would take a person's arms and begin to spin. As we gained speed, their legs would elevate and we would try to raise their bodies until they were parallel to the ground.

Later when we were older we visited the amusement park where rides based on the development of normal acceleration due to rotation were prevalent. The rotating swings were common in many parks. An example of the 48 seat wave swinger rides from Zierer at Gröna Lund, Sweden is shown below. The swings are suspended from three circles with slightly different diameters, and slightly different chain lengths. As the angular velocity of the rotating top increases, the chain suspended seats move outward and upward to balance the normal acceleration relative to gravitational acceleration.

Looping rollercoasters also rely on speed as the cars execute the loop and the passengers are for a moment suspended at the top of the loop, with only centrifugal (inertial) forces holding them in place.

Today we recognize normal accelerations when driving at high speed on the interstate highway. When approaching a curve we try to check how the curve is banked to determine if we need to reduce speed. Of course no calculation is made, but with experience, we seem to sense the angle of the bank and the speed at which the curve can be safely negotiated.

A swing ride, Sweden, where centrifugal forces lift the swings and drive them outward. Gravitational forces maintain equilibrium.

4.1 INTRODUCTION

In Chapter 2 and 3, we dealt with one, two and three dimensional motion of particles. In this chapter we will consider rigid bodies instead of particles and restrict the body's motion to the x-y plane. The study of planar motion of rigid bodies is often divided into four topics including:

1. Rectilinear translation or straight line motion from one location to another without rotation of the rigid body.
2. Curvilinear translation or motion along a curved planar path from one location to another without rigid body rotation.
3. Rotation of the rigid body about a fixed axis.
4. General planar motion where a rigid body moves along a curvilinear path and rotates in the process.

Let's initially consider general planar motion of rigid bodies.

4.2 GENERAL PLANAR MOTION OF RIGID BODIES

To treat the general case of plane motion of a rigid body, we employ a set of reference coordinates OXYZ, as shown in Fig. 4.1. We introduce a second moving set of coordinates Qxyz, which locates a specific plane on or within the rigid body. This plane may translate or rotate, but its distance from the XY plane must remain constant for planar motion. If the body rotates, it does so about point Q.

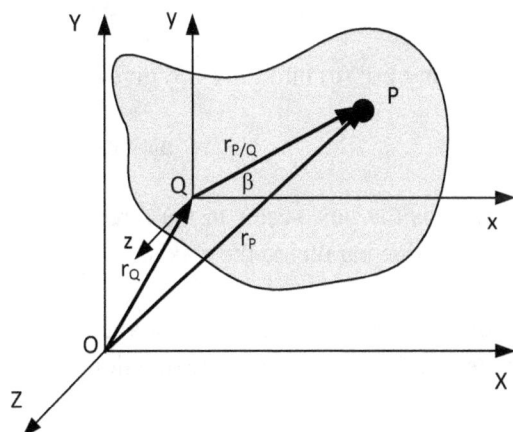

Fig. 4.1 Qxy represents a plane in a rigid body with a fixed distance from the OXY plane.

We locate a point P on the Qxy plane, some distance from the origin Q. We also construct lines r_P and r_Q from the origin O of the reference frame to points P and Q, respectively. The radial line $r_{P/Q}$ that locates point P in the Qxyz coordinate system makes an angle β with the x axis. It is evident from the geometry in Fig. 4.1 that:

$$\mathbf{r_P} = \mathbf{r_Q} + \mathbf{r_{P/Q}} \tag{4.1}$$

Note that the vector $\mathbf{r_{P/Q}}$ is measured from point Q to point P.

We differentiate Eq. (4.1) with respect to time to obtain the velocity $\mathbf{v_P}$ as:

$$\mathbf{v_P} = d\mathbf{r_P}/dt = d\mathbf{r_Q}/dt + d\mathbf{r_{P/Q}}/dt \tag{a}$$

The velocity $\mathbf{v_P} = d\mathbf{r_P}/dt$ is due to translation and rotation of the rigid body. The term $v_Q = d\mathbf{r_Q}/dt$ is the velocity of point Q due to translation of the rigid body. The term $d\mathbf{r_{P/Q}}/dt$ is the velocity of point P due to the angular velocity ω of the body about point Q.

Because the body is rigid $r_{P/Q}$ is constant in magnitude, but the line $r_{P/Q}$ may change direction due to rotation about point Q. We write the velocity due to rotation of line $r_{P/Q}$ as:

$$d\mathbf{r_{P/Q}}/dt = r_{P/Q}\, \omega\, \mathbf{u_\beta} \tag{b}$$

where ω is the angular velocity of the line $r_{P/Q}$ and $\mathbf{u_\beta}$ is a unit vector perpendicular to $\mathbf{r_{P/Q}}$ in the direction of increasing β.

Substituting Eq. (b) into Eq. (a) yields:

$$\mathbf{v_P} = \mathbf{v_Q} + d\mathbf{r_P}/dt = \mathbf{v_Q} + r_{P/Q}\, \omega\, \mathbf{u_\beta} \tag{4.2}$$

To obtain the acceleration $\mathbf{a_P}$, we differentiate $\mathbf{v_P}$ with respect to time to obtain:

$$\mathbf{a_P} = d\mathbf{v_P}/dt = d/dt\,[\mathbf{v_Q} + r_{P/Q}\, \omega\, \mathbf{u_\beta}] \tag{4.3}$$

From Eq. (3.38) it is clear that:

$$d/dt\,[r_{P/Q}\, \omega\, \mathbf{u_\beta}] = r_{P/Q}\,[\omega(d/dt)\, \mathbf{u_\beta} + \alpha\, \mathbf{u_\beta}] = r_{P/Q}\,[-\omega^2\, \mathbf{u_r} + \alpha\, \mathbf{u_\beta}] \tag{c}$$

Substituting Eq. (c) into Eq. (4.3) gives:

$$\mathbf{a_P} = d\mathbf{v_P}/dt = \mathbf{a_Q} - r_{P/Q}\, \omega^2\, \mathbf{u_r} + r_{P/Q}\, \alpha\, \mathbf{u_\beta} \tag{4.4}$$

where $\mathbf{u_r}$ is the unit vector in the direction from Q to P and α is the angular acceleration of the line $r_{P/Q}$ about any line parallel to the z axis.

Another approach involving vector algebra and cross products is interesting given that the vector relations for the angular velocity and acceleration are:

$$\omega = d\beta/dt \; \mathbf{k} = \omega \, \mathbf{k} \qquad \text{and} \qquad \boldsymbol{\alpha} = d^2\beta/dt^2 \; \mathbf{k} = d\omega/dt \; \mathbf{k} = \alpha \, \mathbf{k} \qquad (4.5)$$

Note that the term $r_{P/Q} \, \omega \, \mathbf{u}_\beta$ in Eq. (4.2) is equivalent to the cross product $\boldsymbol{\omega} \times \mathbf{r}_{P/Q}$. The cross product $\boldsymbol{\omega} \times \mathbf{r}_{P/Q}$ gives the magnitude $r_{P/Q} \, \omega$ and the direction $\mathbf{u}_\beta$. Using the cross product representation enables us to rewrite Eq. (4.2) as:

$$\mathbf{v}_P = \mathbf{v}_Q + \boldsymbol{\omega} \times \mathbf{r}_{P/Q} \qquad (4.6)$$

Similarly note that in Eq. (4.4) the term $\rho\alpha \, \mathbf{u}_\beta$ is the same as $\boldsymbol{\alpha} \times \boldsymbol{\rho}$ and the term $-\rho\omega^2 \, \mathbf{u}_r$ may be replaced by $\boldsymbol{\omega} \times (\boldsymbol{\omega} \times \boldsymbol{\rho})$. Making these changes to Eq. (4.4) gives:

$$\mathbf{a}_P = \mathbf{a}_Q + \boldsymbol{\omega} \times (\boldsymbol{\omega} \times \mathbf{r}_{P/Q}) + \boldsymbol{\alpha} \times \mathbf{r}_{P/Q} \qquad (4.7)$$

Finally, we note that Eqs. (4.6) and (4.7) can be rewritten as:

$$\mathbf{v}_P = \mathbf{v}_Q + \mathbf{v}_{P/Q} \qquad (4.8)$$

$$\mathbf{a}_P = \mathbf{a}_Q + \mathbf{a}_{P/Q} \qquad (4.9)$$

4.3 TRANSLATION OF RIGID BODIES

4.3.1 Rectilinear Translation

Let's consider a rectangular (rigid) body that is undergoing rectilinear translation, as depicted in Fig. 4.2. Two views are shown corresponding to positions at times t_1 and t_2 The dashed lines represent the track of the motion of points P and Q.

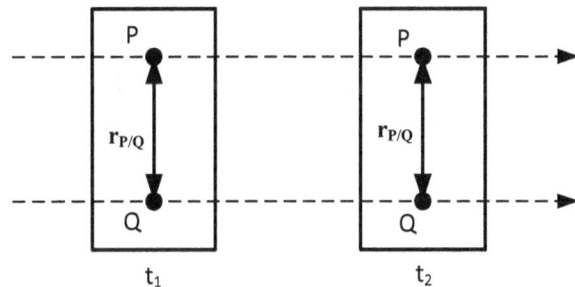

Fig. 4.2 A rigid body undergoing rectilinear translation at time t_1 and t_2.

The rigid body does not rotate during the translation and the distance $\mathbf{r}_{P/Q}$ between point P and Q is a constant (fixed). It is evident from Eqs. (4.8) and (4.9) that:

$$\mathbf{v}_P = \mathbf{v}_Q \qquad\qquad \mathbf{a}_P = \mathbf{a}_Q \qquad (4.10)$$

4.3.2 Curvilinear Translation

Let's consider a rectangular (rigid) body that is undergoing curvilinear translation, as depicted in Fig. 4.3. Two views are shown corresponding to the body's positions at time t_1 and t_2. The dashed lines represent the track of the motion of points P and Q.

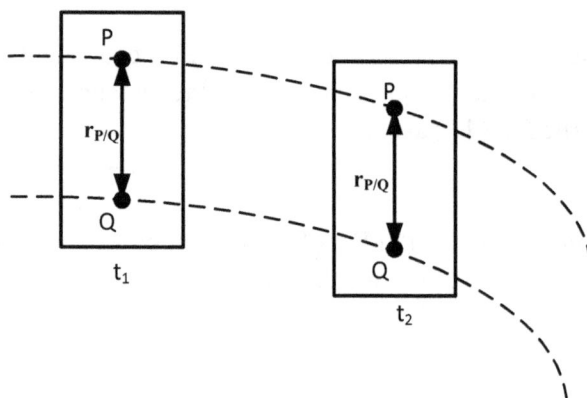

Fig. 4.3 A rigid body undergoing curvilinear
translation at time t_1 and t_2.

The rigid body does not rotate during the translation and the distance $r_{P/Q}$ between point P and Q is a constant (fixed). It is evident from Eqs. (4.8) and (4.9) that:

$$\mathbf{v}_P = \mathbf{v}_Q \qquad\qquad \mathbf{a}_P = \mathbf{a}_Q \qquad\qquad (4.11)$$

Equation (4.11) is identical with Eq. (4.10) indicating that all points in a rigid body undergoing translation have the same velocity and acceleration. Therefore a rigid body in either rectilinear or curvilinear translation can be treated as a particle. The equations developed in Chapters 2 and 3 for particle motion can be applied to rectilinear and curvilinear motion.

4.4 FIXED AXIS ROTATION OF RIGID BODIES

4.4.1 Velocity and Acceleration Due to Fixed Axis Rotation

Consider the rigid body in Fig. 4.1 and fix the axis of rotation by locking the origin Q in place and rotating the body about a line through Q that is parallel to the Z axis. This constraint implies that $\mathbf{v}_Q$ in Eq. (4.2) and $\mathbf{a}_Q$ in Eq. (4.4) vanish. Hence, we may write:

$$\mathbf{v}_P = + r_{P/Q}\, \omega\, \mathbf{u}_\beta \qquad\qquad (4.12)$$

$$\mathbf{a}_P = - r_{P/Q}\, \omega^2\, \mathbf{u}_r + r_{P/Q}\, \alpha\, \mathbf{u}_\beta \qquad\qquad (4.13)$$

$$\mathbf{v}_P = \omega \times \mathbf{r}_{P/Q} \qquad\qquad (4.14)$$

$$\mathbf{a}_P = \omega \times (\omega \times \mathbf{r}_{P/Q}) + \alpha \times \mathbf{r}_{P/Q} \qquad\qquad (4.15)$$

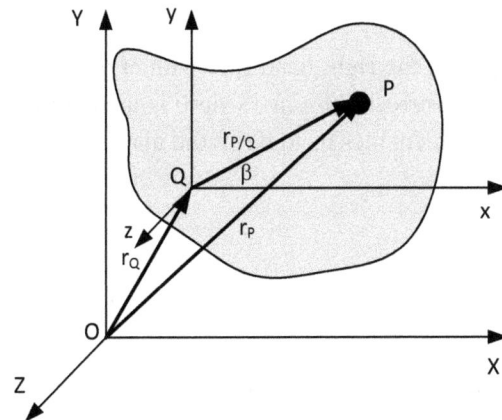

Fig. 4.1 This illustration is a repeat of Fig. 4.1. In this figure consider point Q fixed and the body rotating about this point.

EXAMPLE 4.1

A rigid body is rotating about a fixed axis with an angular velocity ω with its vector along the axis of rotation, as shown in Fig E4.1. A point P located within the body sweeps out a circular path with a radius r, which is given by $r_P \sin \theta$. Determine the velocity $\mathbf{v_P}$ of the point P.

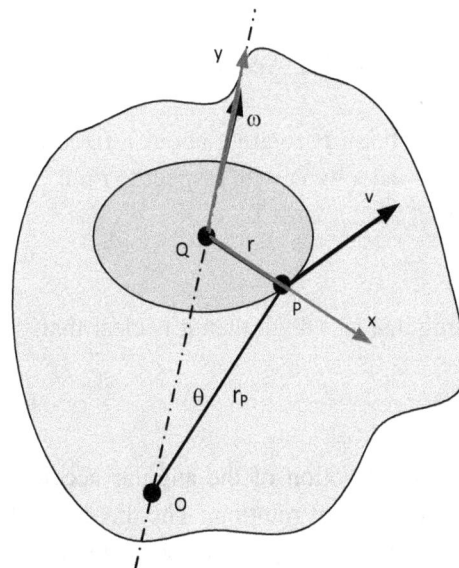

Fig. E4.1 A rigid body rotating about a fixed axis O-Q with an angular velocity ω.

Solution:

The point P travels around a circular path with a radius r where $r = r_P \sin \theta$. The circle inset in Fig. E4.1 represents the plane of motion. The velocity of the point P is:

$$v_P = r \, \omega \tag{a}$$

or

$$v_P = (r_P \sin \theta) \, \omega \tag{b}$$

The velocity v_P is tangent to the circle and its direction is shown in Fig. E4.1. The velocity of point P can also be obtained from the cross product as:

$$\mathbf{v_P} = \boldsymbol{\omega} \times \mathbf{r_P} \qquad\qquad (c)$$

Applying the right hand rule with ω positive (counterclockwise), shows that the velocity direction in Fig. E4.1 is correct. Noting $r_P \sin\theta$ is the component in the x direction and $r_P \sin\theta$ is the component in the y direction enables us to write the matrix for the cross product as:

$$\begin{pmatrix} \mathbf{i} & \mathbf{j} & \mathbf{k} \\ 0 & \omega & 0 \\ r_p\sin\theta & r_p\cos\theta & 0 \end{pmatrix} \qquad\qquad (d)$$

The magnitude of the cross product is:

$$v_P = -\omega\,(r_P \sin\theta) \qquad\qquad (e)$$

We employed the matrix to determine the magnitude and the right hand rule to determine direction.

4.4.2 Angular Acceleration of a Point P Due to Fixed Axis Rotation

When a body is rotating about a fixed axis, the angular acceleration α is obtained by differentiating the angular velocity ω with respect to time:

$$\alpha = d\omega/dt \qquad\qquad (4.16)$$

Noting that $\omega = d\theta/dt$ then it is clear that:

$$\alpha = d^2\theta/dt^2 \qquad\qquad (4.17)$$

The line of action of the angular acceleration coincides with the line of action of the angular velocity along the axis of rotation. The direction of α is positive if ω is increasing and negative if ω is decreasing. With decreasing angular velocity α is called angular deceleration.

Recognize that $dt = d\omega/\alpha$ and $dt = d\theta/\omega$ then:

$$\alpha\,d\theta = \omega\,d\omega \qquad\qquad (4.18)$$

Special Case of Constant Angular Acceleration α_k.

If the angular acceleration is a constant α_k, we can integrate Eqs. (4.16), (4.17) and (4.18) to obtain:

$$d\theta/dt = \omega = \alpha_k\,t + \omega_0 \qquad\qquad (4.19)$$

where ω_0 is the constant of integration equal to the angular velocity at $t = 0$.

Integrate Eq. (4.19) to obtain:

$$\theta = \tfrac{1}{2}\,\alpha_k t^2 + \omega_0\, t + \theta_0 \tag{4.20}$$

where θ_0 is the constant of integration equal to the angular displacement at $t = 0$.

Integrate Eq. (4.18) to obtain:

$$\alpha_k\,(\theta - \theta_0) = \tfrac{1}{2}\,(\omega^2 - \omega_0{}^2) \qquad\qquad \omega^2 = \omega_0{}^2 + 2\,\alpha_k\,(\theta - \theta_0) \tag{4.21}$$

where ω_0 and θ_0 are initial values at $t = 0$.

EXAMPLE 4.2

A point P is located to the rim of a flywheel that is 4 ft in diameter. The flywheel undergoes a uniform increase in angular velocity from 3 to 12 radians/s over a time period of 40 s. Determine the tangential and normal acceleration of the point P at $t = 40$ s and the distance it travels during the 40 s period.

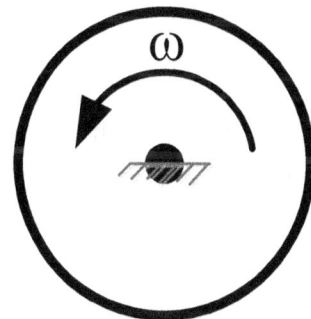

Fig. E4.2 A flywheel, 4 ft. in diameter, with an angular velocity ω that is increasing at a uniform rate with time.

Solution:

Because the angular velocity increases uniformly the angular acceleration α is a constant and equal to:

$$\alpha_k = \Delta\omega/\Delta t = (12 - 3)/40 = 0.225 \text{ rad/s} \tag{a}$$

The tangential acceleration is given by $a_t = \alpha_k\, r$ as:

$$a_t = \alpha_k\, r = 0.225\,(2) = 0.45 \text{ ft/s}^2 \tag{b}$$

The normal acceleration is given by $a_n = \omega^2\, r$ as:

$$a_n = \omega^2\, r = (12)^2\,(2) = 288 \text{ ft/s}^2 \tag{c}$$

This value is the maximum normal acceleration that occurs at 40 s.

The distance travelled by point P is $s = r\,\theta$ where θ is given by Eq. (4.20):

$$\theta = \theta_0 + \omega_0 t + \frac{1}{2}\,\alpha_k\,t^2 \tag{d}$$

Note $\theta_0 = 0$ and $\omega_0 = 3$ radians/s.

$$\theta = \omega_0 t + \frac{1}{2}\,\alpha_k\,t^2 = 3(40) + \frac{1}{2}\,(0.225)(40)^2 = 300 \text{ radians} \tag{e}$$

and

$$s = r\,\theta = 2(300) = 600 \text{ ft} \tag{f}$$

EXAMPLE 4.3

A flywheel is rotating with an initial angular velocity $\omega_0 = 6$ radians/s in the counterclockwise direction, as shown in Fig. E4.3. The flywheel is undergoing a constant angular acceleration $\alpha_k = 2$ rad/s^2. Determine the number of revolutions the flywheel makes before reaching the angular velocity $\omega = 12$ radians/s. Also determine the time required to achieve this angular velocity.

Fig. E4.3 A flywheel rotating at an initial angular velocity $\omega = 6$ rad/s and accelerating at a constant rate of 2 rad/s^2.

Solution:

The time required to achieve $\omega = 12$ rad/s is given by:

$$t = \Delta\omega/\alpha_k = (12 - 6)/2 = 3 \text{ s} \tag{a}$$

The number of rotations under constant angular acceleration is given by Eq. (4.20) as:

$$\theta = \theta_0 + \omega_0 t + \frac{1}{2}\,\alpha\,t^2 = 0 + 6\,(3) + \frac{1}{2}\,(2)(3)^2 = 27 \text{ rad} \tag{b}$$

or

$$\text{Revs} = 27/(2\pi) = 4.297 \tag{c}$$

EXAMPLE 4.4

A hoist consisting of a wheel and a cable is lifting a weight W a distance of 2 m, as shown in Fig. E4.4. The wheel is initially at rest, but when activated undergoes a constant angular acceleration α for 1.5 s and achieves an angular velocity ω = 15 RPM. It maintains this angular velocity until the weight is lifted a distance of 2 m. Determine the total time required for the weight to be lifted.

Fig. E4.4 A hoist lifting a weight W
a distance of 2 m.

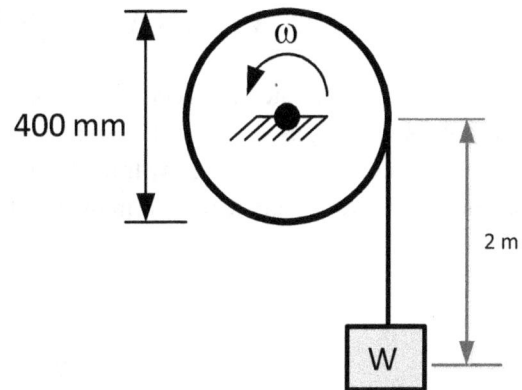

Fig. E4.4 shows 400 mm, ω, 2 m, W.

Solution:

The distance of the lift is related to the angular displacement θ by:

$$\theta = s/r = 2{,}000 \text{ mm}/200\text{mm} = 10 \text{ rad} \qquad (a)$$

Converting 15 RPM into rad/s gives the angular velocity ω at t = 1.5 s:

$$\omega = 15 \text{ rev/min} \times \text{min}/60 \text{ s} \times 2\pi/\text{rev} = \pi/2 \text{ rad/s} \qquad (b)$$

The angular acceleration α_k is determined by:

$$\alpha_k = \Delta\omega/\Delta t = (\pi/2)/(1.5) = \pi/3 \text{ rad/s}^2 \qquad (c)$$

For the time increment from 0 to 1.5 s, the rotational displacement $\theta_{t=1.5}$ is determined from Eq. (4.20):

$$\theta_{t=1.5} = \theta_0 + \omega_0 t + \tfrac{1}{2} \alpha_k t^2 = 0 + 0 + \tfrac{1}{2} (\pi/3)(1.5)^2 = 1.178 \text{ rad} \qquad (d)$$

Because the wheel was initially at rest $\theta_0 = \omega_0 = 0$

The angular displacement after the wheel reaches an angular velocity of ω = π/2 is given by Eqs. (a) and (d) as:

$$\theta_2 = 10 - 1.178 = 8.822 \text{ rad} \qquad (e)$$

The time required to achieve the angular displacement θ_2 is determined from Eq. (b) and Eq. (e) as:

$$t_2 = \theta_2/\omega = 8.822(2)/\pi = 5.616 \text{ s} \qquad\qquad \text{(f)}$$

The total time to lift the weight is the time $t_1 = 1.5$ s to accelerate and the time t_2 to achieve the required angular displacement:

$$t = t_1 + t_2 = 1.5 + 5.616 = 7.116 \text{ s} \qquad\qquad \text{(g)}$$

4.5 RELATIVE VELOCITY OF TWO POINTS ON A RIGID BODY

In Section 4.2, we treated the general case of plane motion of a rigid body by employing a set of reference coordinates OXYZ, as shown in Fig. 4.1. We introduce a second moving set of coordinates Qxyz, which locates a specific plane on or within the rigid body. This plane may translate and/or rotate, but its distance from the XY plane must remain constant. We then derived the following equations:

$$\mathbf{r}_P = \mathbf{r}_Q + \mathbf{r}_{P/Q} \qquad\qquad (4.1)$$

and

$$\mathbf{v}_P = \mathbf{v}_Q + \mathbf{v}_{P/Q} \qquad\qquad (4.8)$$

$$\mathbf{a}_P = \mathbf{a}_Q + \mathbf{a}_{P/Q} \qquad\qquad (4.9)$$

To enhance the discussion of relative motion, we will introduce Fig. 4.4, which is similar to Fig. 4.1; however, it shows a bar instead of a general shape.

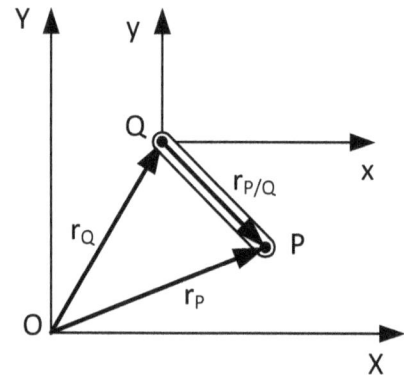

Fig. 4.4 One end of the bar is fixed at point Q of the coordinate system Qxy, which may translate. The coordinate system OXY is fixed in space.

The reference frame OXY is fixed in space, but $\mathbf{r}_Q$ is changing due to the translation of point Q. The vector $\mathbf{r}_{P/Q}$ is changing due to the translation of point Q, the rotation of the bar about the origin Q and the movement of point P due to this rotation. We express these changes as:

$$\Delta\mathbf{r}_P = \Delta\mathbf{r}_Q + \Delta\mathbf{r}_{P/Q} \qquad\qquad \text{(a)}$$

Dividing all three terms in Eq. (a) by Δt and taking the limit of each term as Δt goes to zero yields:

$$\mathbf{v}_P = \mathbf{v}_Q + \mathbf{v}_{P/Q} \qquad\qquad (4.8)$$

This result confirms Eq. (4.8) that was derived previously. Note that $\mathbf{v_P}$ and $\mathbf{v_Q}$ are measured relative to the fixed coordinate system OXY and are absolute velocities independent of the rotation of the bar. However, $\mathbf{v_{P/Q}}$ is due to the rotation of the bar. Accordingly we may write:

$$\mathbf{v_{P/Q}} = r_{P/Q}\, \mathbf{d\theta/dt} = r_{P/Q}\, \boldsymbol{\omega} \tag{4.22}$$

where ω is the angular velocity of the bar about a z axis through point Q.

To better visualize the terms in Eqs. (4.8) and (4.22) consider the diagram, shown in Fig. 4.5.

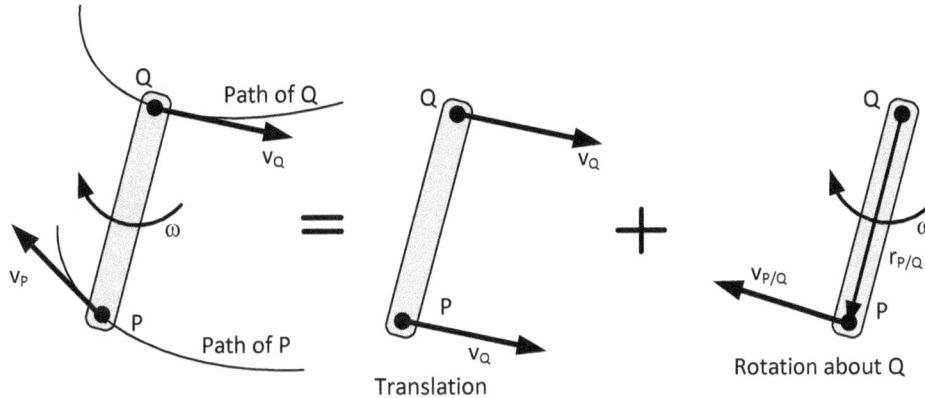

Fig. 4.5 Graphical depiction of Eq. (4.8), which shows the motion is a translation of the bar plus its rotation about point Q.

For the rotation term note that $\mathbf{v_{P/Q}} = r_{P/Q}\, \boldsymbol{\omega}$ and the vector diagram showing the vector addition given in Eq. (4.8) is presented in Fig. 4.6.

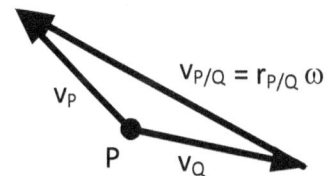

Fig. 4.6 The vector diagram for $\mathbf{v_P} = \mathbf{v_Q} + r_{P/Q}\, \boldsymbol{\omega}$

The term $v_{P/Q}$ is due to only rotation. Hence, we can write it as a cross product shown below:

$$\mathbf{v_{P/Q}} = \boldsymbol{\omega} \times \mathbf{r_{P/Q}} \tag{4.23}$$

Substituting Eq. (4.23) into Eq. (4.8) yields:

$$\mathbf{v_P} = \mathbf{v_Q} + \boldsymbol{\omega} \times \mathbf{r_{P/Q}} \tag{4.24}$$

The relations developed above are useful in solving problems with bar linkages or problems with bars attached to wheels with pin joints. They are also employed to address problems of wheels rolling along a plane surface, because the contact point of the wheel with the surface can be considered a point about which the wheel is rotating. This point is called the instantaneous center. This application of Eq. (4.8) is illustrated in Fig. 4.7. In this example the linear velocity $\mathbf{v_P}$ of the wheel is given by:

$$\mathbf{v_P} = \boldsymbol{\omega} \times \mathbf{r_{P/Q}}$$

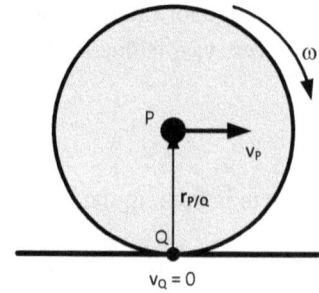

Fig 4.7 Point Q is the instantaneous center for a wheel rotating with an angular velocity ω.

EXAMPLE 4.5

A wheel with a cylindrical drum attached is rolled by a cable wrapped about the cylindrical drum. If the cable is pulled with a velocity of 6 m/s, determine the velocity at point P located on top of the wheel, as shown in Fig E4.5. The wheel rolls without slipping at point Q.

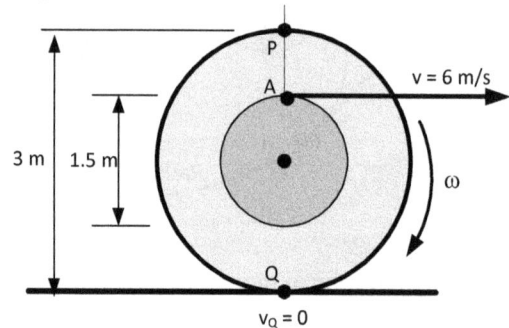

Fig. E4.5 A cable wrapped around the drum is pulled with a velocity of 6 m/s.

Solution:

The velocity of point A is given by:

$$\mathbf{v_A} = \mathbf{v_Q} + r_{A/Q}\,\boldsymbol{\omega} = 0 + \omega\,[(D-d)/2] = 6 \text{ m/s} \qquad\text{(a)}$$

Solving Eq. (a) for ω gives:

$$\omega = \frac{2v_A}{D+d} = \left(\frac{2(6)}{3+1.5}\right) = 2.667 \text{ rad / s clockwise} \qquad\text{(b)}$$

The velocity of point P is:

$$\mathbf{v_P} = \mathbf{v_Q} + r_{P/Q}\,\boldsymbol{\omega} = 0 + D\,\omega = 3\,(2.667) = 8.000 \text{ m/s} \qquad\text{to the right} \qquad\text{(c)}$$

EXAMPLE 4.6

A two bar linkage arrangement is shown in Fig. E4.6. The bar AB is 0.3 m long and is rotating with an angular velocity $\omega_{AB} = 6$ rad/s in the counterclockwise direction. The bar BC is 0.5 m long and is pinned to bar B on its left end and to a piston on its right end. Bar BC is horizontal at the instant under consideration. The piston at point C is constrained to a channel that is oriented 60^0 to the x axis. Determine the velocity at point C and the angular velocity of bar BC.

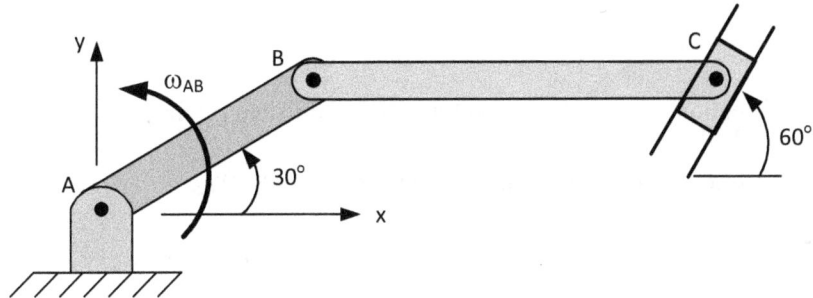

Fig. E4.6 Bar and piston mechanism.

Solution:

Consider the bar A- B as a free body, as shown in Fig E4.6a.

Fig. E4.6a The velocity vector $\mathbf{v_B}$, due to the angular velocity ω_{AB}, is perpendicular to bar A-B.

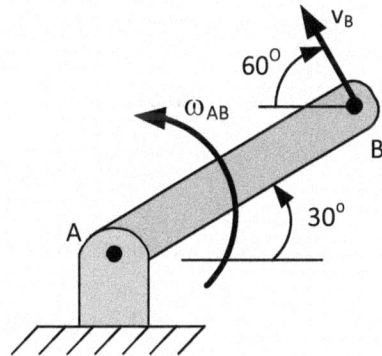

The velocity $\mathbf{v_B}$ is given by:

$$\mathbf{v_B} = \mathbf{v_A} + \omega_{AB} \times \mathbf{r_{B/A}} \qquad (a)$$

Substituting numerical values into Eq. (b) gives:

$$\mathbf{v_B} = 0 + 6\,\mathbf{k} \times [+0.3\cos 30\,\mathbf{i} + 0.3\sin 30\,\mathbf{j}] \qquad (b)$$

$$\mathbf{v_B} = 6\,\mathbf{k} \times [0.2598\,\mathbf{i} + 0.15\,\mathbf{j}] = 1.559\,\mathbf{j} - 0.9\,\mathbf{i} \qquad (c)$$

Writing Eq. (c) in standard format yields:

$$\mathbf{v_B} = -0.9\,\mathbf{i} + 1.559\,\mathbf{j} \qquad (d)$$

Next remove bar BC as a free body, as shown in Fig. E4.6b:

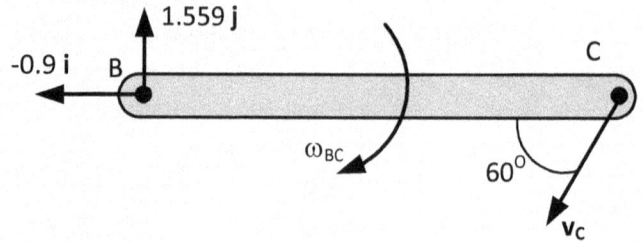

Fig. E4.6b Bar BC is horizontal at this instant
of time.

The direction of the velocity vector $\mathbf{v_C}$ is established by the constraint on the motion of the piston.

Next we write the equation for the velocity at point C as:

$$\mathbf{v_C} = \mathbf{v_B} + \boldsymbol{\omega}_{BC} \times \mathbf{r}_{C/B} \qquad (e)$$

Substituting the results from Eq. (d) into Eq. (e) yields:

$$v_C(-\cos 60^\circ\, \mathbf{i} - \sin 60^\circ\, \mathbf{j}) = -0.9\, \mathbf{i} + 1.559\, \mathbf{j} + (-\omega_{BC})\mathbf{k} \times (0.5\, \mathbf{i}) \qquad (f)$$

The term $(-\omega_{BC})$ is negative because the rotation bar BC is clockwise.

Expanding Eq. (e) gives:

$$v_C(-0.5\, \mathbf{i} - 0.8660\, \mathbf{j}) = -0.9\, \mathbf{i} + 1.559\, \mathbf{j} - 0.5\, \omega_{BC}\, \mathbf{j} \qquad (f)$$

The coefficient of the $\mathbf{i}$ term gives:

$$0.5\, v_C = 0.9 \qquad \text{and} \qquad v_C = 1.8 \text{ m/s} \qquad (g)$$

The coefficient of the $\mathbf{j}$ term gives:

$$1.8(-0.8660) = 1.559 - 0.5\, \omega_{BC} \qquad (h)$$

Solving Eq. (h) for ω_{BC} yields:

$$\omega_{BC} = 3.118/0.5 = 6.236 \text{ rad/s} \qquad (i)$$

EXAMPLE 4.7

A rod A-B that is 2 m long connects with a piston at position A and to a wheel with a diameter of 1.2 m at position B, in the mechanism shown in Fig. E4.7. The piston moves to the left with a velocity of 5 m/s. Determine the velocity at point B and the angular velocity ω_{AB} of rod A-B.

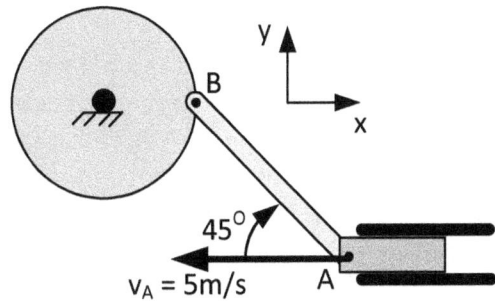

Fig. E4.7 A piston driven mechanism.

Solution:

Begin by removing bar A-B as a free body, as shown in Fig. E4.7a. We have drawn vectors v_A and v_B at the two ends of bar A-B and recognized the geometric constrains that enable us to provide the directions for the vectors.

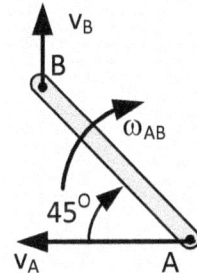

Fig. E4.7a A free body of rod A-B showing the direction of vectors v_A and v_B.

The velocity v_B is given by:

$$v_B = v_A + \omega_{AB} \times r_{B/A} \qquad (a)$$

Substituting numerical values into Eq. (b) gives:

$$j v_B = -5\,i + (-\omega_{AB})\,k \times [-2 \cos 45^O\,i + 2 \sin 45^O\,j] \qquad (b)$$

Note that the ω_{AB} term is negative, because the bar A-B is rotating clockwise.

Equation (b) reduces to:

$$j v_B = -5\,i + \omega_{AB}\,[1.414\,j + 1.414\,i] \qquad (c)$$

The **i** coefficient gives:

$$-5 + 1.414\,\omega_{AB} = 0$$

Then:

$$\omega_{AB} = 5/1.414 = 3.536 \text{ rad/s} \qquad (d)$$

As expected the angular velocity is positive because our choice of its sign was correct in writing Eq. (b).

The **j** coefficient gives:

$$v_B = 1.414\,\omega_{AB} = 1.414(3.536) = 5 \text{ m/s} \qquad \text{upward} \qquad (e)$$

4.6   INSTANTANEOUS CENTER

We have already introduced the concept of instantaneous center in the solution to Exercise 4.5. The concept is important, because we often can recognize a point on a body that is at rest, while the remainder of the body is in motion. A wheel that rolls without slipping is an example, where the point in contact with the ground is at rest relative to the ground. The advantage is that this knowledge enables us to simplify Eq. (4.8) by letting $v_Q = 0$. By placing point Q at an instantaneous center with $v_Q = 0$, Eq. (4.8) reduces to:

$$\mathbf{v_P} = \mathbf{v_{P/Q}} = \boldsymbol{\omega} \times \mathbf{r_{P/Q}} \tag{4.25}$$

With point Q at the instantaneous center, we can place point P at any location of the body and compute the velocity at that point using the cross product $\boldsymbol{\omega} \times \mathbf{r_{P/Q}}$. The angular velocity $\boldsymbol{\omega}$ is usually known and the vector $\mathbf{r_{P/Q}}$ is established from the geometry of the body.

In some cases the instantaneous center is not obvious by inspection; however, if we know the velocities at two points in the body it is easy to determine the location of the instantaneous center. An example showing two velocity vectors $\mathbf{v_P}$ and $\mathbf{v_D}$ is illustrated in Fig. 4.8. The velocity vectors are tangent to arcs that identifies the instantaneous center at point C. The radii $r_{P/C}$ and $r_{D/C}$ make an angle of 90 with the velocity vectors.

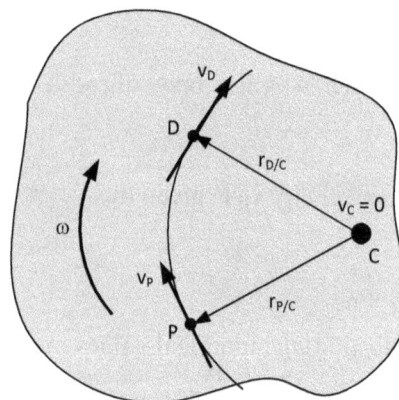

Fig. 4.8 Determining the location of the instantaneous center from velocity vectors at two points in a body.

Another case, where the instantaneous center can be determined, involves two parallel velocity vectors that are pointed in opposite directions, as shown in Fig. 4.9. In this case, it is evident that:

$$r_{P/C} = v_P/\omega \qquad \text{or} \qquad r_{DC} = v_D/\omega \qquad \text{and} \qquad r_{P/C} + r_{D/C} = L$$

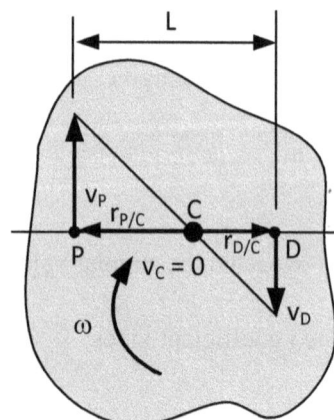

Fig. 4.9 Determining the location of the instantaneous center from two parallel but opposing vectors at two points in a body.

Another case, where the instantaneous center can be determined, involves two parallel velocity vectors that are pointed in the same direction, as shown in Fig. 4.10 In this case, it is evident that the location of the instantaneous center can be determined by using the rule of similar triangles.

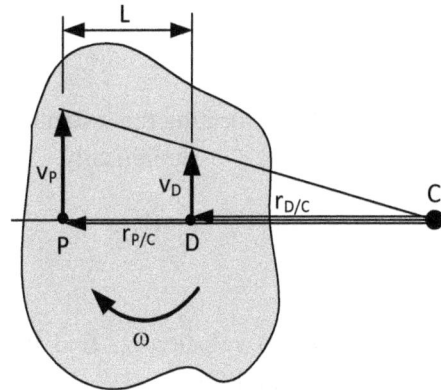

Fig. 4.10 Determining the location of the instantaneous center from two parallel vectors in the same direction.

Finally, if the velocity v_P and the angular velocity ω are known, the radius $r_{P/C}$ establishing instantaneous center along a radial line from v_P can be determined from:

$$r_{P/C} = v_P/\omega$$

EXAMPLE 4.8

A ladder is erected against a wall, as shown in Fig. E4.8. The vertical and horizontal surfaces are slippery and the ladder begins to slide downward with a velocity v_B = 12 ft/s and to the left with an unknown velocity v_A. Determine the angular velocity ω of the ladder and the velocity at its center point C.

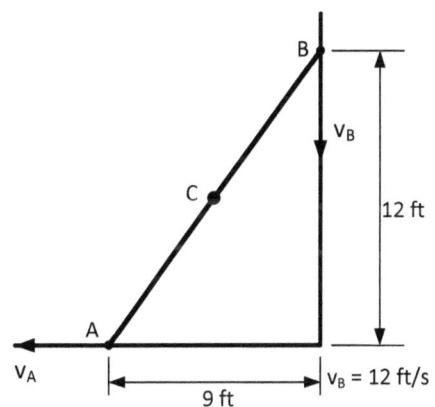

Fig. E4.8 Ladder resting against a wall and supported at its feet, becomes unstable and begins to slide down and to the right.

Solution:

First determine the instantaneous center IC using a graphical approach, as shown in Fig. E4.8a. Note that $r_{B/IC}$ = 9 ft. Then we determine the magnitude of angular acceleration ω_{AB} as:

$$\omega = 12/9 = 4/3 = 1.333 \text{ rad/s} \qquad \text{(a)}$$

Fig. E4.8a Locating the instantaneous center IC graphically.

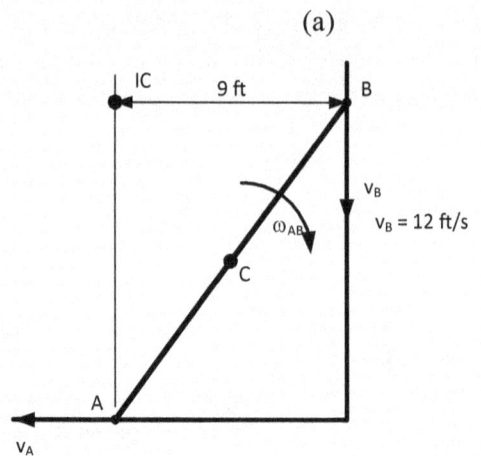

To determine the velocity v_C, find the distance $r_{C/IC} = 7.5$ ft, by using the geometry of the triangles presented in Fig. E4.8b.

Fig. E4.8b Triangles associated with the ladder and the location of the IC.

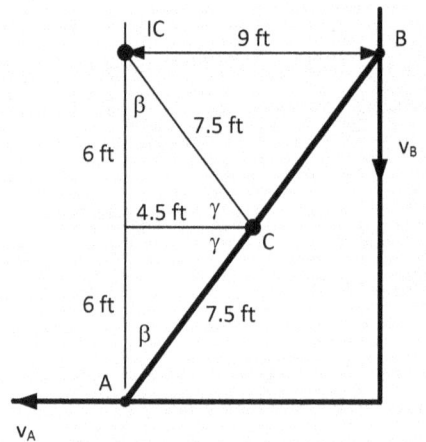

Fig. E4.8c The vector $\mathbf{v}_C$ due to rotation about the IC.

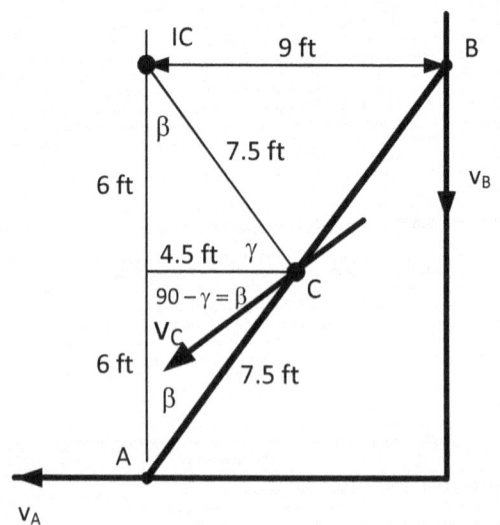

The magnitude of the velocity v_C is given by:

$$v_C = r_{C/IC}\, \omega_{AB} = 7.5(1.333) = 10 \text{ ft/s} \qquad \text{(b)}$$

Let's calculate the angles β and γ from the diagram in Fig. E4.8c as:

$$\gamma = \tan^{-1} 6/4.5 = \tan^{-1} 1.333 = 53.13^O \qquad\qquad \beta = 90^0 - 53.13^0 = 36.87^0 \qquad (c)$$

Next note the vector v_C on the diagram shown in Fig. 4.8c and observe the angle β = 36.87^O that the velocity vector makes with a horizontal line.

EXAMPLE 4.9

The mechanism presented in Fig. E4.9 has 6 ft diameter wheel that drives a 12 ft long connecting rod, which in turn acts on a piston that makes an angle of 60° with the connecting rod. The wheel has a constant angular velocity ω = 9 rad/s. Determine the angular velocity of the connecting rod and the velocity of the piston.

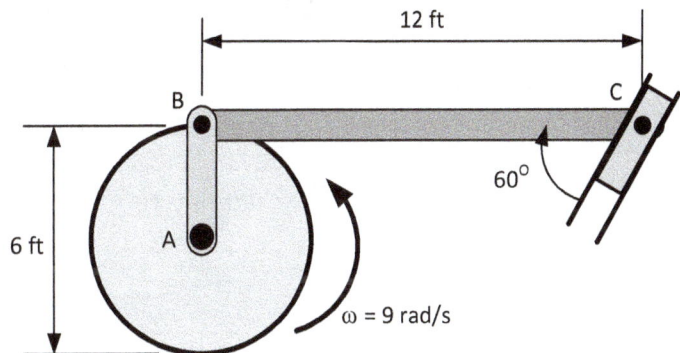

Fig. E4.9 The rotating wheel acts on a connecting rod that in turn drives the piston at point C.

Solution:

Remove the connecting rod as a free body and graphically identify the location of the instantaneous center IC, as shown in Fig. E4.9.

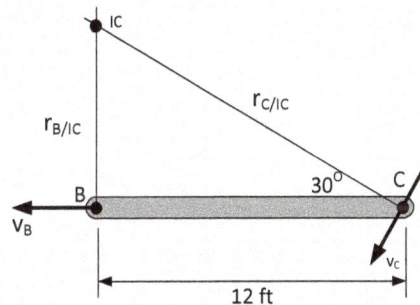

Fig. E4.9 Locating the IC by graphical construction of perpendicular lines.

It is evident from the geometry of the triangle that:

$$r_{C/IC} \cos 30^0 = 12 \qquad and \qquad r_{C/IC} = 12/0.8660 = 13.86 \text{ ft} \qquad (a)$$

Then

$$r_{B/IC} = r_{C/IC} \sin 30^0 = 13.86\,(0.5) = 6.928 \text{ ft} \qquad (b)$$

Because the wheel is rotating about a fixed axis, we may write:

$$v_B = r_{AB}\, \omega = 3(9) = 27 \text{ ft/s} \tag{c}$$

The angular velocity ω_{BC} of the connecting rod B-C is:

$$\omega_{BC} = v_B/r_{B/IC} = 27/6.928 = 3.897 \text{ rad/s} \qquad \text{in the clockwise direction} \tag{d}$$

The velocity of point C is:

$$v_C = r_{C/IC}\, \omega_{BC} = 13.86(3.897) = 54.01 \text{ ft/s} \qquad \text{down and to the left} \tag{e}$$

4.7 RELATIVE ACCELERATION OF TWO POINTS ON A RIGID BODY

We have already considered the general case of acceleration of a rigid body in plane motion in Section 4.2 and derived the equation:

$$\mathbf{a}_P = d\mathbf{v}_P/dt = \mathbf{a}_Q - r_{P/Q}\, \omega^2\, \mathbf{u_r} + r_{P/Q}\, \alpha\, \mathbf{u_\beta} \tag{4.4}$$

Equation (4.4) can be rewritten as:

$$\mathbf{a}_P = \mathbf{a}_Q + (\mathbf{a}_{P/Q})_n + (\mathbf{a}_{P/Q})_t \tag{4.26}$$

The terms $\mathbf{a}_P$ and $\mathbf{a}_Q$ represent acceleration of points P and Q relative to a fixed set of coordinates OXY; whereas, the terms $(\mathbf{a}_{P/Q})_n$ and $(\mathbf{a}_{P/Q})_t$ represent the relative acceleration of point P with respect to the origin associated with the moving coordinate system Qxy. The relative acceleration $\mathbf{a}_{P/Q}$ has been resolved into its normal component $(\mathbf{a}_{P/Q})_n$ and its tangential component $(\mathbf{a}_{P/Q})_t$.

Comparing Eq. (4.26) with Eq. (4.4) shows:

$$(a_{P/Q})_n = r_{P/Q}\, \omega^2 \tag{4.27}$$

and

$$(a_{P/Q})_t = r_{P/Q}\, \alpha \tag{4.28}$$

The direction of the normal component of acceleration is from P to Q, as shown in Fig. 4.1 and the direction of the tangential component is perpendicular to $r_{P/Q}$.

It is useful to rewrite Eq. (4.4) in terms of the normal and tangential components as:

$$\mathbf{a}_P = \mathbf{a}_Q - \omega^2\, \mathbf{r}_{P/Q} + \mathbf{\alpha} \times \mathbf{r}_{P/Q} \tag{4.29}$$

These relations are used when determining the acceleration of pin connected components, where $a_Q = 0$.

EXAMPLE 4.10

A flywheel, fixed at point A, drives a 5 ft long connecting rod B-C, which in turn drives the piston located at point C, as shown in Fig. E4.10. The flywheel, with a diameter of 2.5 ft, is rotating with an angular velocity $\omega_{AB} = 8$ rad/s and is accelerating with $\alpha_{AB} = 6$ rad/s^2 in the clockwise direction. Determine the angular acceleration α_{BC} of the connecting rod and the acceleration at points B and C on the connecting rod.

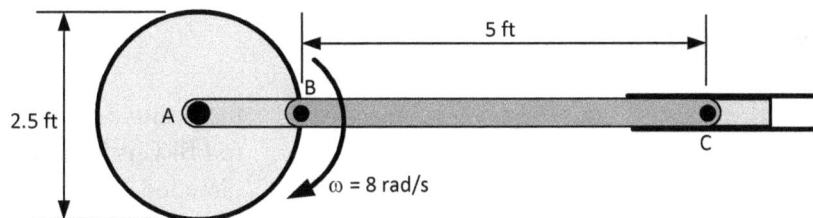

Fig. E4.10 A flywheel driving a connecting rod, which in turn drives a piston. At this instant $v_c = 0$.

Solution:

Prepare a free body diagram of the connecting rod B-C, as illustrated in Fig. E4.10a.

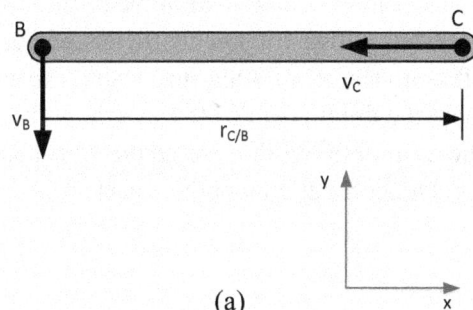

Fig. 4.10a A free body diagram of the connecting rod B-C.

It is evident from the FBD that:

$$r_{C/B} = 5\,\mathbf{i} \qquad\qquad (a)$$

The velocity v_B is given by:

$$v_B = \omega r_{B/A} = 8(1.25) = 10 \text{ ft/s} \qquad \text{downward} \qquad (b)$$

The angular velocity of the connecting rod B-C is:

$\omega_{BC} = v_B/r_{C/B} = 10/(5) = 2$ rad/s   The rod is rotating in the counterclockwise direction   (c)

The acceleration of point B located on the flywheel is determined from Eq. (4.15) as:

$$\mathbf{a_B} = -\omega^2\,\mathbf{r_{B/A}} + \boldsymbol{\alpha} \times \mathbf{r_{B/A}} \qquad\qquad (d)$$

$$\mathbf{a_B} = -(8)^2\,(1.25)\,\mathbf{i} - 6\,\mathbf{k} \times 1.25\mathbf{i} = -80\,\mathbf{i} - 7.5\,\mathbf{j} \qquad\qquad (e)$$

The acceleration of point C located on the piston is determined from Eq. (4.29) as:

$$\mathbf{a_C} = \mathbf{a_B} - \omega_{BC}^2\,\mathbf{r_{C/B}} + \alpha_{BC} \times \mathbf{r_{C/B}} \qquad\qquad (d)$$

Substituting numerical results into Eq., (d) yields:

$$i a_C = -80\,\mathbf{i} - 7.5\,\mathbf{j} - (2)^2(5)\mathbf{i} + \alpha_{BC}\,\mathbf{k} \times 5\,\mathbf{i} \qquad (e)$$

$$i a_C = -80\,\mathbf{i} - 7.5\,\mathbf{j} - (20)\,\mathbf{i} + 5\,\alpha_{BC}\,\mathbf{j} \qquad (f)$$

The **i** coefficient yields:

$$a_C = -100 \text{ ft/s}^2 \qquad \text{the negative sign indicates point C}$$
is accelerating toward the left. (g)

The **j** coefficient yields:

$$\alpha_{BC} = 7.5/5 = 1.5 \text{ rad/s}^2 \qquad \text{the positive sign indicates the angular acceleration of}$$
rod B- C is in the counterclockwise
direction. (h)

EXAMPLE 4.11

A wheel is in contact with a block B, as shown in Fig. E4.11. The contact between the wheel and the block has a high friction coefficient and as such the wheel does not slip when the block B moves to the left at a velocity of 3 m/s and with an acceleration of 1.5 m/s². A drum is fixed to the wheel and a cable about the drum is pulled to the right with a velocity of 6 m/s and with a tangential acceleration of 3 m/s². The radii defining the size of the wheel and the drum are 0.8 m and 0.4 m respectively. Determine the angular acceleration α of the wheel.

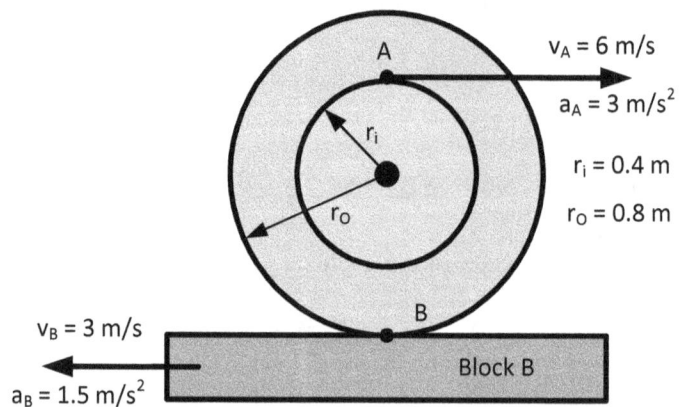

Fig. E4.11 A wheel rotating on block B
without slipping.

$v_A = 6$ m/s

$a_A = 3$ m/s²

$r_i = 0.4$ m

$r_o = 0.8$ m

$v_B = 3$ m/s

$a_B = 1.5$ m/s²

Block B

Solution:

Let's remove the wheel and drum as a free body and draw the velocity vectors at points A and B, as shown in Fig. E4.11a. Connecting a line from the tip of v_A to the tip of v_B, enables us to locate the instantaneous center at point C.

Fig. E4.11a Locating the instantaneous center at point C.

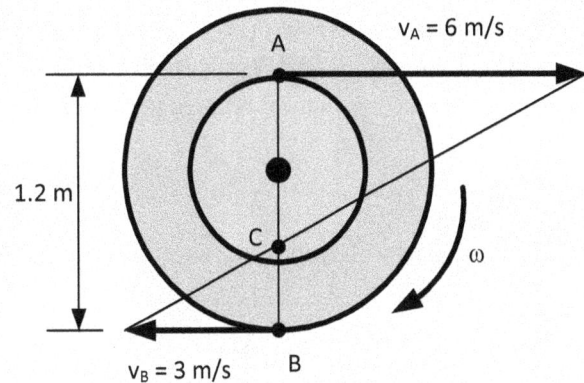

From similar triangles it is evident that:

$$r_{C/B}/3 = r_{A/C}/6 \qquad \text{and} \qquad r_{A/C} = 2\, r_{C/B} \qquad \text{(a)}$$

Also the distance from A to B is:

$$r_{C/B} + r_{A/C} = 1.2 \qquad \text{then} \qquad 3\, r_{C/B} = 1.2 \qquad \text{(b)}$$

$$r_{C/B} = 0.4 \text{ m} \qquad \text{and} \qquad r_{A/C} = 0.8 \text{ m} \qquad \text{(c)}$$

The angular velocity ω of the wheel is given by:

$$\omega = v_A/\, r_{A/C} = 6/0.8 = 7.5 \text{ rad/s} \qquad \text{clockwise} \qquad \text{(d)}$$

The tangential acceleration at point A is given by:

$$\mathbf{a_A} = \mathbf{a_B} + \boldsymbol{\alpha} \times \mathbf{r_{A/C}} \qquad \text{(e)}$$

Substituting for $\mathbf{a_A}$ and $\mathbf{a_B}$ known vector quantities gives:

$$3\,\mathbf{i} = -1.5\,\mathbf{i} - \alpha\,\mathbf{k} \times 0.8\,\mathbf{j} \qquad \text{(f)}$$

Executing the cross product and simplifying Eq. (f) gives:

$$4.5\,\mathbf{i} = 0.8\alpha\,\mathbf{i} \qquad \text{(g)}$$

The coefficients of the $\mathbf{i}$ terms yields:

$$\alpha = 4.5/0.8 = 5.625 \text{ rad/s}^2 \qquad \text{(h)}$$

In writing Eqs. (e) and (f) we have neglected the normal components of acceleration at point A and B. Both of these components contribute only to the coefficient of the $\mathbf{j}$ term and as such do not affect the magnitude of the angular acceleration α of the wheel.

4.8   SUMMARY

We began our discussion of kinematics of rigid bodies undergoing planar motion by defining a position vector $\mathbf{r_P}$ that identifies a point in a body relative to a fixed coordinate system OXY. We then established a moving coordinate system Qxy with a position vector $\mathbf{r_Q}$ that locates Q relative to the coordinates OXY. Finally we located the point P relative to Q with the position vector $\mathbf{r_{P/Q}}$ and expressed $\mathbf{r_P}$ as:

$$\mathbf{r_P} = \mathbf{r_Q} + \mathbf{r_{P/Q}} \tag{4.1}$$

Differentiating $\mathbf{r_P}$ with respect to time yields the velocity vector $\mathbf{v_P}$ as:

$$\mathbf{v_P} = d\mathbf{r_P}/dt = \mathbf{v_Q} + r_{P/Q}\,\omega\,\mathbf{u_\beta} \tag{4.2}$$

Differentiating $\mathbf{v_P}$ with respect to time yields the acceleration vector $\mathbf{a_P}$ as:

$$\mathbf{a_P} = d\mathbf{v_P}/dt = \mathbf{a_Q} - r_{P/Q}\,\omega^2\,\mathbf{u_r} + r_{P/Q}\,\alpha\,\mathbf{u_\beta} \tag{4.4}$$

where $\mathbf{u_r}$ and $\mathbf{u_\beta}$ are unit vectors in the r and β directions, respectively.

The angular velocity $\boldsymbol{\omega}$ and acceleration $\boldsymbol{\alpha}$ are given by:

$$\boldsymbol{\omega} = \omega\,\mathbf{k} \quad \text{and} \quad \boldsymbol{\alpha} = d\omega/dt\,\mathbf{k} = \alpha\,\mathbf{k} \tag{4.5}$$

Equation (4.2) may also be written as:

$$\mathbf{v_P} = \mathbf{v_Q} + \boldsymbol{\omega} \times \mathbf{r_{P/Q}} \tag{4.6}$$

Equation (4.4) may also be written as:

$$\mathbf{a_P} = \mathbf{a_Q} + \boldsymbol{\omega} \times (\boldsymbol{\omega} \times \mathbf{r_{P/Q}}) + \boldsymbol{\alpha} \times \mathbf{r_{P/Q}} \tag{4.7}$$

Finally, we note that Eqs. (4.6) and (4.7) can be rewritten as:

$$\mathbf{v_P} = \mathbf{v_Q} + \mathbf{v_{P/Q}} \tag{4.8}$$

$$\mathbf{a_P} = \mathbf{a_Q} + \mathbf{a_{P/Q}} \tag{4.9}$$

We treated fixed axis rotation, by locking the origin Q in place and rotate the body about a line through Q that was parallel to the Z axis. This constraint implies that $\mathbf{v_Q}$ in Eq. (4.2) and $\mathbf{a_Q}$ in Eq. (4.4) vanish. Hence:

$$\mathbf{v_P} = \boldsymbol{\omega} \times \mathbf{r_{P/Q}} \tag{4.14}$$

$$\mathbf{a_P} = \boldsymbol{\omega} \times (\boldsymbol{\omega} \times \mathbf{r_{P/Q}}) + \boldsymbol{\alpha} \times \mathbf{r_{P/Q}} \tag{4.15}$$

For a body is rotating about a fixed axis, the angular acceleration α is obtained by:

$$\alpha = d\omega/dt \qquad (4.16)$$

or

$$\alpha = d^2\theta/dt^2 \qquad (4.17)$$

The line of action of the angular acceleration coincides with the line of action of the angular velocity along the axis of rotation. The direction of α is positive if ω is increasing and negative if ω is decreasing.

$$\alpha\, d\theta = \omega\, d\omega \qquad (4.18)$$

If the angular acceleration is a constant α_k, then we can integrate Eqs. (4.16), (4.17) and (4.18) to obtain:

$$\omega = \alpha_k\, t + \omega_0 \qquad (4.19)$$

$$\theta = \tfrac{1}{2}\, \alpha_k t^2 + \omega_0\, t + \theta_0 \qquad (4.20$$

$$\alpha_k\, (\theta - \theta_0) = \tfrac{1}{2}\, (\omega^2 - \omega_0^2)$$

or $\qquad\qquad\qquad\qquad\qquad\qquad\qquad\qquad\qquad (4.21)$

$$\omega^2 = \omega_0^2 + 2\, \alpha_k\, (\theta - \theta_0)$$

We then treated the relative velocity of two points on a rigid body and showed:

$$\mathbf{v}_{P/Q} = r_{P/Q}\, \mathbf{d\theta/dt} = r_{P/Q}\, \boldsymbol{\omega} \qquad (4.22)$$

Because $\mathbf{v}_{P/Q}$ is due only to rotation, we write it as a cross product:

$$\mathbf{v}_{P/Q} = \boldsymbol{\omega} \times \mathbf{r}_{P/Q} \qquad (4.23)$$

or

$$\mathbf{v}_P = \mathbf{v}_Q + \boldsymbol{\omega} \times \mathbf{r}_{P/Q} \qquad (4.24)$$

We identified the significant advantage of identifying the instantaneous center IC of a body in motion. By placing the origin Q at the IC enables us to simplify Eq. (4.8) by letting $\mathbf{v}_Q = 0$ and write:

$$\mathbf{v}_P = \mathbf{v}_{P/Q} = \boldsymbol{\omega} \times \mathbf{r}_{P/Q} \qquad (4.25)$$

With point Q at the instantaneous center, we can place point P at any location of the body and compute the velocity at that point using the cross product $\boldsymbol{\omega} \times \mathbf{r}_{P/Q}$. The angular velocity $\boldsymbol{\omega}$ is usually known and the vector $\mathbf{r}_{P/Q}$ is established from the geometry of the body.

Equation (4.4) can be rewritten as:

$$\mathbf{a}_P = \mathbf{a}_Q + (\mathbf{a}_{P/Q})_n + (\mathbf{a}_{P/Q})_t \qquad (4.26)$$

where

$$(a_{P/Q})_n = r_{P/Q}\, \omega^2 \qquad (4.27)$$

and

$$(a_{P/Q})_t = r_{P/Q}\, \alpha \qquad (4.28)$$

The direction of the normal component of acceleration is from P to Q and the direction of the tangential component is perpendicular to $r_{P/Q}$. It is useful to rewrite Eq. (4.4) as:

$$\mathbf{a}_P = \mathbf{a}_Q - \omega^2\, \mathbf{r}_{P/Q} + \boldsymbol{\alpha} \times \mathbf{r}_{P/Q} \tag{4.29}$$

CHAPTER 5

KINETICS OF PARTICLE MOTION

INTRODUCTION

In this chapter you will learn how to apply Newton's second law to characterize particle motion. This topic is the first of several that covers Kinetics. Initially you will deal with translation that is produced when forces are applied to a particle. We will ask you to study examples that are presented using both the SI and the U S Customary system of units.

You will be applying Newton's second law using three different coordinate systems that include: rectangle coordinates, normal and tangential coordinates and radial and transverse coordinates. You will select a coordinate system that best matches the motion of the particle along either a linear or a curvilinear path. The directions of particle motion associated with the tangential and normal coordinates are established with tangential and normal unit vectors $\mathbf{u_t}$ and $\mathbf{u_n}$ shown to the right:

Tangential and normal unit vectors $\mathbf{u_t}$ and $\mathbf{u_n}$.

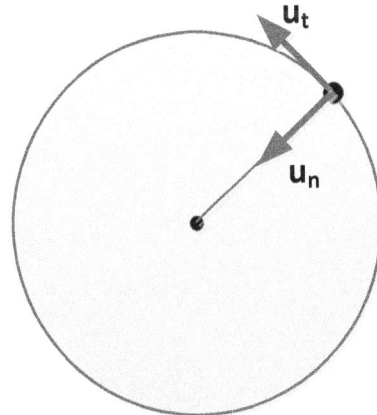

The directions of particle motion associated with the transverse and radial coordinates are established with transverse and radial unit vectors $\mathbf{u_r}$ and $\mathbf{u_T}$ that are illustrated in the figure to the right:

Transverse and radial unit vectors $\mathbf{u_T}$ and $\mathbf{u_r}$.

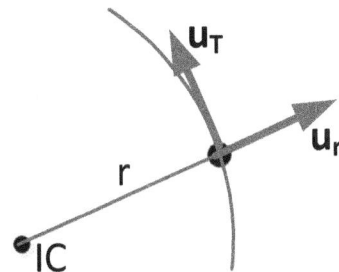

When solving problems involving any of the coordinate systems, you first draw a free body diagram of the particle after selecting the appropriate coordinate system. You then write equations for ΣF based on this diagram. Then you compute the magnitude of the acceleration components based on knowledge of the curvilinear path of the particle using the appropriate equations. Combining the ΣF equations with the components of acceleration yields two or three relationships that you solve to determine the unknown forces acting on the particle.

CONCEPT PROBLEM

As an analysist for a state department of transportation, you are assigned the task of determining the banking angle for constructing curves on two-lane highways. The banking angle is to be determined for vehicles of any weight traveling around curves at a constant speed of 60 MPH. You are to develop a table showing the banking angle in degrees for radii of curvature of 200, 400, 600, 800 and 1,000 ft. Assume that the banking angle is sufficient to eliminate the need for lateral friction forces to develop. If the banking angle is not sufficient lateral forces would generate friction forces that would oppose these lateral forces and prevent the vehicle from skidding. The solution or perhaps your driving experience will show that the sharper the curve (smaller radius of curvature) the larger angle of the banking required to prevent drivers from spinning out and skidding off of the highway. Speed also enters into the automobiles stability in negotiating curves. The higher speeds require higher banking angles.

To begin prepare a free body diagram of a vehicle moving around a curve with a radius of curvature ρ and a banking angle θ, as shown below.

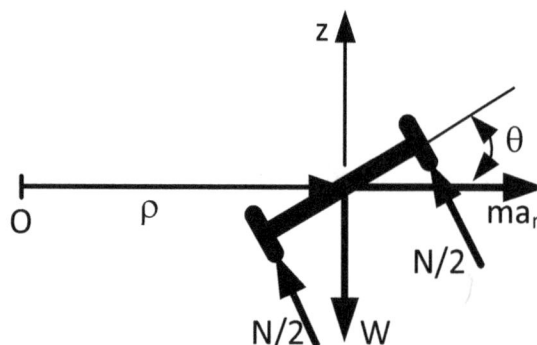

Forces acting on vehicle traversing banked curve in a highway.

From the free body diagram you write:

$$\Sigma F_z = 0 \qquad 2(N/2) \cos \theta - W = 0 \qquad\qquad \text{(a)}$$

Note that $W = mg$ and Eq. (a) yields:

$$N \cos \theta = mg \qquad\qquad \text{(b)}$$

Consider $\Sigma F_n = 0$ and write the equilibrium equation to obtain:

$$\Sigma F_n = ma_n - 2 (N/2) \sin \theta = 0 \qquad\qquad \text{(c)}$$

Recall $a_n = v^2/\rho$ and Eq. (c) reduces to:

$$N \sin \theta = mv^2/\rho \qquad\qquad \text{(d)}$$

Combining results from Eq. (b) and Eq. (d) yields:

$$\tan \theta = v^2/\rho g \tag{e}$$

Conversion of 60 MPH to ft/s gives v = 88 ft/s.

Substituting numerical results into Eq. (e) for v and g gives:

$$\tan \theta = (240.7/\rho) \tag{f}$$

Results for the banking angle θ in Eq. (f) for the specified values of ρ are shown in the table below:

Radius of Curvature (ft)	Banking Angle, Degrees
200	50.28
400	31.04
600	21.86
800	16.75
1,000	13.53

The results in the table clearly show that the banking angle required to prevent an automobile from losing control decreases as the radius of curvature of the highway increases. Speed also enters into the stability of automobiles in negotiating curves as indicated by $\tan \theta = v^2/\rho g$. Note the velocity is term is squared, which indicates the very significant influence of speed.

DISCUSSION

The analysis performed to determine banking angle for curves is one example of government's efforts to make our highways safer. More work remains to be done to reduce the number of accidents that occur each year on our highways. Each year about 40,000 people are killed on our highways and more than another 4 million are injured and require medical treatment. Preliminary estimates from the National Safety Council (NSC) indicate motor vehicle deaths dipped in 2017, with a death toll of 40,100 down slightly from 40,327 deaths in 2016. In 2017, about 4.57 million people were injured seriously enough to require medical attention with a cost of $413.8 billion to those injured.

Deborah A.P. Hersman, the President of the National Safety Council (NSC) stated: "The price we are paying for mobility is 40,000 lives each year. This is a stark reminder that our complacency is killing us. The only acceptable number is zero; we need to mobilize a full court press to improve roadway safety. Complacency is killing us."

Over the last decade, new cars have been equipped with electronic stability control systems to prevent skids, rear and side view cameras to prevent collisions and more airbags to protect occupants from injury in both front and side collisions. Hundreds of millions of dollars have been spent on campaigns to remind the public of the dangers of drunken driving, failing to buckle up and texting while driving. Despite all these efforts, many Americans are dying on roads and highways each year. Moreover the sudden and sharp increase in accidents in the past decade has caused concern among safety advocates.

Part of the increase is believed to stem from the improving economy, which has enabled us to drive more miles for both work and pleasure. But safety advocates state that the improved economy explains only part of the increase, because the number of deaths as a percentage of miles driven is also increasing. They also point to data showing an increase in distracted driving. While cars and phones often offer advanced voice controls and other features intended to keep drivers' eyes on the road, apps like Facebook, Google Maps, others have created new temptations that drivers find hard to resist. Drivers' texting while driving is a significant distraction, because the driver often looks at the smartphone for more than a second or two. How far will an auto travel in 4 seconds at 60 MPH? What is the probability of encountering an obstacle or another auto during those 4 seconds and travelling 352 ft at 88 ft/s? Recently the NSC called for a ban on all use of smartphones by drivers, even if they use hands-free calling or messaging. The council is also advocating mandatory motorcycle helmet laws and ignition locks that prevent repeat drunken drivers from operating their vehicles while impaired.

Government officials and safety advocates believe that some of the reasons for the increase in fatalities have been caused by more lenient enforcement of seatbelt, drunken driving and speeding regulations by authorities. Also reluctance by lawmakers to pass more restrictive measures is another reason. A patchwork of state laws leaves many states in which drivers can choose not to buckle up, with little probability of being ticketed. Only 18 states have laws requiring seatbelts for both front and rear occupants and categorize not wearing them as a primary offense. In 15 states, failure to wear a seatbelt in front seats is only a secondary offense that implies that drivers cannot be ticketed unless they are stopped for other violations.

About half of all traffic fatalities involve unbelted occupants, and almost a third involves drivers who were impaired by drugs or alcohol, according the National Highway Traffic Safety Administration (NHTSA). In some states, highway speed limits are increasing. In recent years, Texas has increased speed limits to 85 miles per hour in some rural areas. About 1,500 miles of roads in Texas have a limit of 75 miles per hour or higher.

Recently the NHTSA began a new initiative called the Road to Zero to eliminate traffic fatalities within 30 years. This initiative relies on the expected improvement in safe operation due to widespread use of autonomous vehicles. However, other experts believe that more needs to be accomplished on basic road safety issues. Ralph Nader has indicated that the way to mitigate the rise in deaths is with a wide range of the nuts-and-bolts measures, not self-driving cars. He has also stated that traffic deaths could be reduced simply by requiring rear-seat occupants to wear seatbelts, installing more cameras to catch speeding drivers without a police presence and tightening regulations on heavy trucking.

Do you text and drive? Do you use your cell phone in discussions with friends while you drive? Can you maintain you concentration on the road ahead while in a conversation? How many miles have you driven since your last accident? How many miles have you driven since your last ticket? Do you use a seat belt when driving? Do you use a seat belt when sitting in the back seat? Have you ever been injured in an accident? Did that experience change your driving habits?

MORE DISCUSSION

Several years ago I was returning home from a visit to the University of Maryland. I was driving south on I-81 approaching an intersection somewhere in Virginia. I was in the right hand lane and had slowed down to about 65 MPH considering exiting the interstate highway to purchase gasoline. Suddenly a sedan cut in front of me. The driver (he) must have been travelling at a velocity of at least 90 MPH, perhaps more. When the driver turned left to straighten out into the right hand lane his rear wheels lost traction and he spun out. He wound up in the ditch along the side of I-81 sliding backwards. No one was injured and his car was not damaged, because of the construction of the ditch enabled the car to slide freely until it stopped.

This driver spun out on a level road with no curve in sight. Clearly, the driver created his own curve as he straightened out into the right hand lane after cutting in sharply to make his pass. Excessive speed was a factor in causing the spin. However, another factor was the weight distribution in the sedan. In today's automobiles, the internal combustion engine, transmission, three pumps and the alternator are located in the front and over the front wheels. Much of the weight of the car is due to these components and most braking is due to the disc brakes on the front wheels. As a consequence, the weight over the rear wheels is relatively small and the frictional grip that the rear wheels make with the highway surface is small. Spin out from either a sharp right or left hand turn on a level highway is easy to accomplish if you want to live dangerously. Speed is also a factor as the radial force that causes the spin increases as the square of the velocity. We developed the equations that show the lateral force as function of velocity and curvature of the path of an automobile in the Concept Problem.

5.1 NEWTON'S EQUATION OF MOTION

In our first experience with kinetics, we will consider a particle rather than a rigid body, because dealing with a particle is less involved. With a particle, we do not have to consider moments that produce rotation of a rigid body. Particle motion involves only translation, which is controlled by Newton's second law that was introduced in Chapter 1.

$$\Sigma \mathbf{F} = m \frac{d\mathbf{v}}{dt} = m\mathbf{a} \tag{5.1}$$

where $\Sigma \mathbf{F}$ is the vector sum of all the forces acting on the particle which has a mass m. The acceleration **a** is a vector quantity. The motion is referenced to an inertial coordinate system OXYZ, which is fixed in space.

Newton's second and third laws are discussed in Section 1.3.2 and we recommend that you review these concepts. Also Newton's gravitational law is described in Section 1.4. Because of its importance, it is repeated below:

$$F = m\, g = W \tag{5.2}$$

and $$g = GM / R_e^2 \tag{5.3}$$

where F is the force acting on a body with a mass m. If the body is Earth bound, g is the gravitational constant and W is the weight of the body. The gravitational constant g is related to G (the universal gravitational constant), the mass of the earth (M), and the radius of the Earth R_e.

In the International System of Units (SI), the mass is expressed in kilograms (kg) and the unit for force is newton (N). A force of 1N imparts an acceleration of 1 m/s^2 to a mass of 1 kg [i. e. 1 N = (1 kg) (1 m/s^2)]. For dimension homogeneity, the newton N is equivalent to $(kg\text{-}m)/s^2$. In the SI system, the **gravitation constant** g = 9.807 m/s^2 and the weight of a mass of 1 kg is W = mg = (1kg)(9.807 m/s^2) = 9.807 N

In the U. S. Customary System, the mass is expressed in slugs. A force of 1 lb will impart an acceleration of 1 ft/s^2 to a mass of 1 slug [i. e. 1 lb = (1 slug)(1 ft/s^2)]. For dimension homogeneity, the slug is equivalent to $(lb\text{-}s^2)/ft$. In the U. S. Customary System the gravitation constant g = 32.17 ft/s^2. With this value for g, the mass of a body weighing 32.17 lb is m = W/g = 32.17 lb/(32.17 ft/s^2) = 1 slug.

EXAMPLE 5.1

A particle weighing 60 N is subjected to a force F = 10 N, determine the particle's mass and its acceleration.

Solution:

The mass of the particle is:

$$m = \frac{W}{g} = \frac{60(N\text{-}s^2)}{9.807m} = 6.118\,kg \tag{a}$$

From Eq. (5.1) we write:

$$a = \frac{F}{m} = \frac{10N(9.807m)}{60(N\text{-}s^2)} = 1.635\,m/s^2 \tag{b}$$

EXAMPLE 5.2

A particle weighing 120 lb is subjected to a force F = 35 lb., determine the particles mass and its acceleration.

Solution:

$$m = \frac{W}{g} = \frac{120(lb\text{-}s^2)}{32.17ft} = 3.730\,slug \tag{a}$$

From Eq. (5.1) we write:

$$a = \frac{F}{m} = \frac{35lb(32.17ft)}{120(lb\text{-}s^2)} = 9.383\,ft/s^2 \tag{b}$$

5.2   RECTANGULAR COORDINATES

In this section and the following two sections, we express the components of force and acceleration in different coordinate systems relative to an inertial reference frame Oxyz. Different coordinate systems are useful in the efficient solution of problems, where the path of the particle conforms to a particular coordinate system. For the rectangular coordinates, we write the equation of motion in a vector format as:

$$\Sigma \mathbf{F} = m\mathbf{a} \tag{5.1}$$

or

$$\Sigma F_x \mathbf{i} + \Sigma F_y \mathbf{j} + \Sigma F_z \mathbf{k} = m(a_x \mathbf{i} + a_y \mathbf{j} + a_z \mathbf{k}) \tag{5.4}$$

The coefficients of the **i**, **j** and **k** terms yield the scalar versions of the equations of motion in rectangular coordinates as:

$$\Sigma F_x = ma_x \qquad \Sigma F_y = ma_y \qquad \Sigma F_z = ma_z \tag{5.5}$$

If the motion of the particle is restricted to the x-y plane, only the x and y components of the scalar equations of motion apply.

EXAMPLE 5.3

A box that weighs 400 lb is moved along a level floor by applying a force F, as shown in Fig. E5.3. The force $F = 190\ t^2 + 320$ lbs. The static and dynamic coefficients of friction between the box and the floor are $\mu_s = 0.25$ and $\mu_d = 0.21$, respectively. Determine the velocity of the box after 3.0 s.

Fig. E5.3 A force F applied to a box to move it along a level floor.

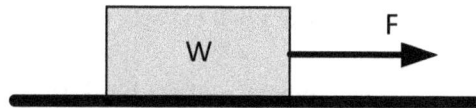

Solution:

Prepare a free body diagram showing the box and forces acting upon it and a reference frame, as shown in Fig. E5.3a.

Fig. E5.3a A free body diagram of the block with a rectangular coordinate system.

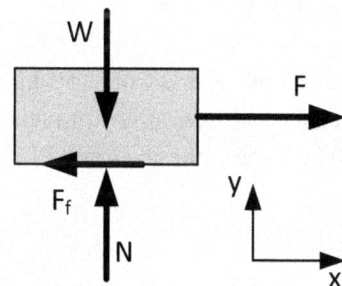

First consider forces in the y direction and write:

$$\Sigma F_y = N - W = 0 \qquad \text{and} \qquad N = W = 400 \text{ lb} \tag{a}$$

Determine the static friction force F_f as:

$$F_f = \mu_s N = 0.25(400) = 100 \text{ lb} \tag{b}$$

Compare the forces in the x direction at time $t = 0$ and note:

$$F = 320 \text{ lb} > F_f = 100 \text{ lb} \tag{c}$$

Because $F > F_f$, the block begins to move at $t = 0$ and the friction force reduces to:

$$F_f = \mu_d N = 0.21(400) = 84 \text{ lb} \tag{d}$$

Using Eq. (5.5) we write the equation of motion as:

$$\Sigma F_x = 190\, t^2 + 320 - 84 = (400/32.17)\, a_x \tag{d}$$

or

$$190\, t^2 + 236 = 12.43\, a_x \qquad \text{or}$$

$$a_x = 15.29 t^2 + 18.99$$

$$a_x = dv/dt = 15.29\, t^2 + 18.99 \tag{e}$$

Integrating Eq. (e) gives:

$$v = 15.29/3\, t^3 + 18.99\, t + C \tag{f}$$

But $C = 0$, because $v = 0$ at $t = 0$.

At $t = 3$ s, the velocity v of the block is:

$$v = (15.29/3)(3)^3 + 18.99\,(3) = 137.61 + 56.97 = 194.6 \text{ ft/s} \tag{g}$$

EXAMPLE 5.4

A block with a mass m = 50 kg is initially at rest, as shown in Fig E5.4. The bock is moving with a velocity v = 4 m/s, after it travels a distance of 5 m in the positive x direction. The coefficient of friction $\mu = 0.3$. Determine the force P.

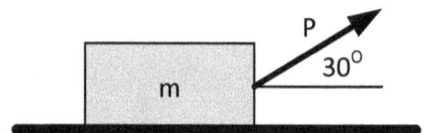

Fig. E5.4 A block moving under the action of a constant force P.

Solution:

Prepare a free body diagram with a coordinate frame, as shown in Fig. E5.4a.

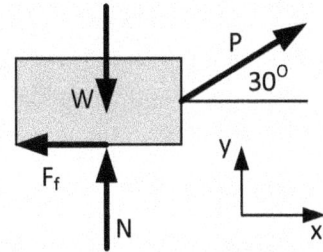

Fig. E5.4a A free body diagram of the block.

First consider forces in the y direction and write:

$$\Sigma F_y = N - W + P \sin 30^O = 0 \quad \text{and} \quad N = W - P/2 \tag{a}$$

Determine the friction force F_f as:

$$F_f = \mu_d N = 0.3(W - P/2) \tag{b}$$

Using Eq. (5.5), we write the equation of motion as:

$$\Sigma F_x = P \cos 30^O - 0.3 W + 0.15 P = 50 a_x \tag{c}$$

$$W = mg = 50 (9.807) = 490.4 \text{ N} \tag{d}$$

Substitute Eq. (d) into Eq. (c) and simplify to obtain:

$$0.0203 P - 2.942 = a_x \tag{e}$$

Examination of Eq. (e) shows that the acceleration is a constant with respect to time. For a body in constant acceleration, we recall Eq. (2.14) and write:

$$v^2 = 2 a_x x \tag{f}$$

Note that the initial values of v and x vanished in Eq. (2.14) because $v_0 = x_0 = 0$ at $t = 0$.

Substituting numerical values into Eq. (f) yields:

$$(4)^2 = 2(5)a_x \qquad\qquad a_x = 1.6 \text{ m/s}^2 \tag{g}$$

Substituting Eq. (g) into Eq. (e) yields:

$$0.203 P - 2.942 = 1.6 \tag{h}$$

or

$$P = 223.7 \text{ N} \tag{i}$$

5.3   NORMAL AND TANGENTIAL COORDINATES

In Section 5.2, we employed rectangular coordinates when dealing with particles traveling along straight lines in the x-y plane. However, if a particle is moving along a curvilinear path in the x-y plane, we often employ normal and tangential components in writing the equation of motion. In this case, we write Eq. (5.1) as:

$$\Sigma F_t \, \mathbf{u_t} + \Sigma F_n \, \mathbf{u_n} = ma_t + ma_n \qquad (5.6)$$

where $\mathbf{u_t}$ and $\mathbf{u_n}$ are unit vectors in the tangential and normal directions respectively.

Recall from Chapter 3, Section 3.3 that:

$$a_t = dv/dt \qquad (5.7)$$

$$a_n = v^2/\rho \qquad (5.8)$$

where ρ is the radius of curvature for the curvilinear path and the direction of the normal component of acceleration is toward the origin for ρ.

It is evident from Eq. (5.6) that the scalar equations of motion in tangential and normal components are:

$$\Sigma F_t = ma_t \qquad \text{and} \qquad \Sigma F_n = ma_n \qquad (5.9)$$

EXAMPLE 5.5

A pendulum, shown in Fig. E5.5, has a mass m = 25 kg and at the 60° position it has a velocity of 5 m/s. The massless rod supporting the mass has a length of 3 m. Determine the force exerted on the rod by the mass of the pendulum and the tangential acceleration at this instant in time.

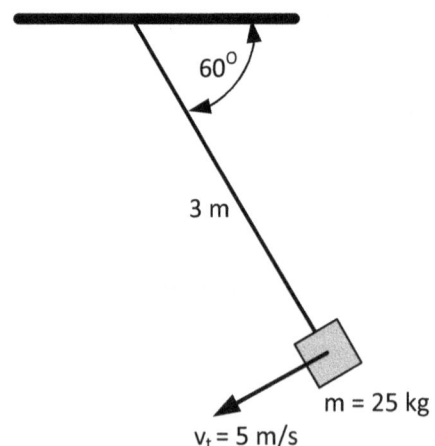

Fig. E5.5 A pendulum swinging due to the gravitational force.

Solution:

Prepare a free body diagram of the pendulum mass, as shown in Fig. E5.5a.

Fig. E 5.5a A free body diagram of the pendulum mass.

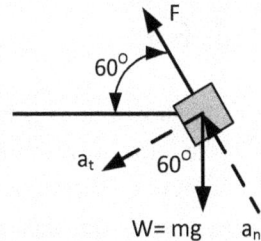

Write the scalar equations of motion for the n and t directions as:

$$\Sigma F_t = mg \sin 30^\circ = ma_t \qquad (a)$$

and

$$\Sigma F_n = F - mg \cos 30^\circ = ma_n = mv^2/\rho \qquad (b)$$

From Eq. (a) we find:

$$a_t = g \sin 30^\circ = 4.904 \text{ m/s}^2 \qquad (c)$$

Substituting into Eq. (b) yields:

$$F = m[g \cos 30^\circ + (5)^2/3] \qquad (d)$$

Evaluating Eq. (d) with m = 25 kg gives:

$$F = 420.7 \text{ N} \qquad (e)$$

5.4   RADIAL AND TRANSVERSE COORDINATES

When considering a particle moving along a curvilinear path in space, we often employ cylindrical coordinates (radial and transverse) and write Eq. (5.1) as:

$$\Sigma F_r \mathbf{u_r} + \Sigma F_T \mathbf{u_T} + \Sigma F_z \mathbf{k} = ma_r + ma_T + ma_z \qquad (5.10)$$

where $\mathbf{u_r}$ and $\mathbf{u_T}$ and $\mathbf{k}$ are unit vectors in the radial, transverse and vertical directions respectively. Recall Eq. (3.45) and note the components of acceleration in the radial and transverse directions:

$$a_r = [d^2r/dt^2 - r(d\theta/dt)^2] \qquad (5.11)$$

$$a_T = [r \, d^2\theta/dt^2 + 2(d\theta/dt)(dr/dt)] \qquad (5.12)$$

where r and θ define the position of the particle as a function of time on the r-T plane.

It is evident from Eq. (5.10) that the scalar equations of motion in cylindrical coordinates are:

$$\Sigma F_r = ma_r \qquad \Sigma F_T = ma_T \qquad \Sigma F_z = ma_z \qquad (5.13)$$

When solving problems involving cylindrical coordinates, we construct a free body diagram of the particle and write equations for ΣF based on the diagram. We then use Eqs. (5.11) and (5.12) to establish the relations for the components of the acceleration based on geometric factors.

EXAMPLE 5.6

The mechanism, shown in Fig. E5.6, is used to lift a cylindrical weight of 10 lb vertically up the side of the block. The arm is rotating with a constant angular velocity $\omega = 1.2$ rad/s. The smooth vertical wall along which the smooth cylinder slides has a friction coefficient $\mu = 0$. Determine the forces acting on the cylinder.

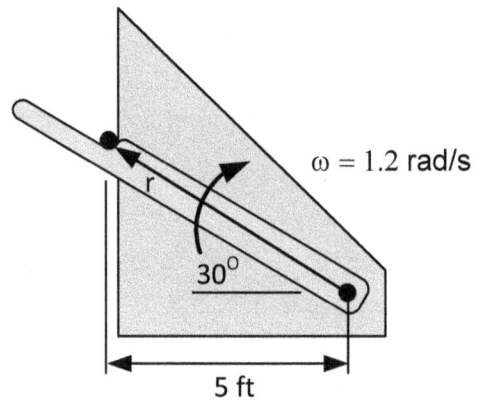

$\omega = 1.2$ rad/s

30°

5 ft

Fig. E5.6 The bar lifts the cylinder up along the smooth
vertical wall.

Solution:

Prepare a free body diagram of the cylinder as shown in Fig. E5.6a.

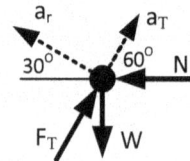

Fig. E5.6a A free body diagram of the weight W.

Write the equations for the forces on the radial and transverse directions.

$$\Sigma F_r: \quad N \cos 30^O - W \sin 30^O = ma_r \tag{a}$$

$$\Sigma F_T: \quad F_T - W \cos 30^O - N \sin 30^O = ma_T \tag{b}$$

Next use the geometry of the mechanism to determine the acceleration components a_r and a_T.

Note: $\qquad\qquad r = 5 \sec 30^O = 5(1.1547) = 5.774$ ft $\qquad\qquad$ (c)

$$dr/dt = 5 \sec \theta \tan \theta \, (d\theta/dt) = 5 \sec \theta \tan \theta \,(1.2) = 6 \sec \theta \tan \theta \tag{d}$$

where θ is the angle through which the bar is rotating. The positive sign for $d\theta/dt$ is due the fact that r is increasing with time and dr/dt is positive.

$$\theta = 30^O \qquad dr/dt = 6(1.1547)(0.5773) = 4.0 \text{ ft/s} \tag{e}$$

$$d^2r/dt^2 = 6[\sec \theta \sec^2 \theta \, (d\theta/dt) + \tan \theta^2 \sec \theta \, (d\theta/dt)]$$
$$= 6[\sec \theta \sec^2 \theta \,(1.2) + \tan \theta^2 \sec \theta \,(1.2)]$$
$$= 7.2[\sec^3 \theta + \tan \theta^2 \sec \theta \,] \tag{f}$$

$\theta = 30^{\circ}$ $\qquad d^2r/dt^2 = 7.2[(1.1547)^3 + (0.5773)^2 (1.1547)] = 13.856 \text{ ft/s}^2$ (g)

Substituting results for r, dr/dt and d^2r/dt^2 into Eq. (5.11) yields:

$$a_r = [d^2r/dt^2 - r(d\theta/dt)^2] = 13.856 - (5.774)(1.2)^2 = 5.541 \text{ ft/s}^2 \qquad \text{(h)}$$

Substituting results for r, dr/dt and d^2r/dt^2 into Eq. (5.12) yields:

$$a_T = [r \, d^2\theta/dt^2 + 2(d\theta/dt)(dr/dt)] = 0 + 2(4)(1.2) = 9.6 \text{ ft/s}^2 \qquad \text{(i)}$$

Substituting the results from Eqs. (h) and (i) into Eqs. (a) and (b) gives:

$\qquad \Sigma F_r$: $\qquad$ N (0.8660) – 10 (0.5) = (10/32.17)(5.541)

and $\qquad$ $\qquad$ N = 6.722/0.8660 = 7.762 lb $\qquad\qquad\qquad\qquad\qquad$ (j)

$\qquad \Sigma F_T$: $\qquad$ F_T – 10(0.8660) – 7.762 (0.5) = (10/32.17)(9.6)

$\qquad\qquad\qquad$ $F_T = 2.984 + 8.660 + 3.881 = 15.53$ lb $\qquad\qquad\qquad$ (k)

EXAMPLE 5.7

The mechanism, shown in Fig. E5.7, includes a slotted plate with vertical orientation and a slotted arm. The slotted arm rotates clockwise and drives a pin along the slot that is cut into the plate. Initially the arm is at $\theta = 0^{\circ}$ and it is rotating clockwise with a constant angular velocity of 3 rad/s. As the arm rotates and the pin moves in the slot, the path it follows is defined by r = 0.6 cos 2θ, where r is measured in m. The pin has a mass of 0.2 kg. Determine the force driving the pin, assuming all of the surfaces are smooth and the friction coefficient $\mu = 0$.

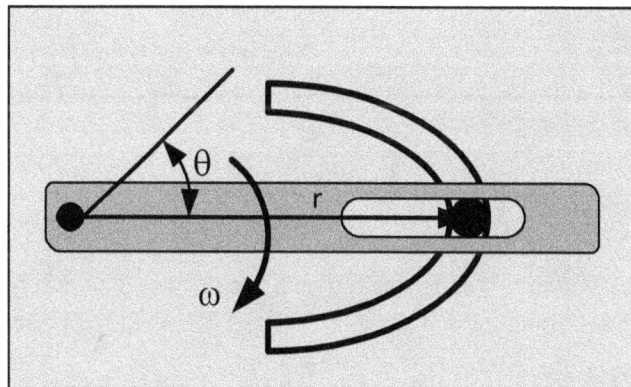

Fig. E5.7 Slotted arm driving pin along a path
formed in a slotted plate.

Solution:

Prepare a free body diagram of the pin as shown in Fig. E5.7a.

Fig. E5.7a A free body diagram of the weight W.

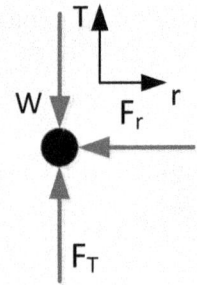

Write the equation for the forces in the transverse directions.

$$\Sigma F_T : \qquad\qquad F_T - W = ma_T \qquad\qquad\qquad (a)$$

Next use the geometry of the mechanism to determine the acceleration component a_T.

$$r = 0.6 \cos 2\theta$$

At $\theta = 0$ $\qquad\qquad\qquad r = 0.6$ m $\qquad\qquad\qquad\qquad (b)$

$$dr/dt = -2(0.6) \sin 2\theta \; d\theta/dt = -1.2(-3) \sin 2\theta = 3.6 \sin 2\theta$$

Note $d\theta/dt = -3$ rad/s because ω is in the negative direction in Fig. E5.7.

At $\theta = 0$ $\qquad\qquad\qquad dr/dt = 0 \qquad\qquad\qquad\qquad (c)$

$$d^2r/dt^2 = 3.6(2) \cos 2\theta \; (d\theta/dt) = 7.2 \, (-3) \cos 2\theta = -21.6 \cos 2\theta$$

At $\theta = 0$ $\qquad\qquad\qquad d^2r/dt^2 = -21.6$ m/s^2 $\qquad\qquad (d)$

$$a_T = [r \, d^2\theta/dt^2 + 2(d\theta/dt)(dr/dt)] = 0.6(0) + 2(-3)(0) = 0 \qquad (e)$$

Recall ΣF_T and note that $\qquad F_T - W = ma_T = 0 \qquad\qquad\qquad (f)$

$$F_T = W = mg = 0.2(9.807) = 19.61 \text{ N} \qquad\qquad (g)$$

EXAMPLE 5.8

A cylindrical rod is rotating on a horizontal plane, as shown in Fig. E5.8. A linear bearing weighing 5 lb slides along the rod without friction. The position parameters r and θ are given by:

$$r = 3t^2 - t^3 \qquad\qquad \text{and } \theta = 2t^2 \qquad\qquad \text{with r given in ft and } \theta \text{ in radians.}$$

Determine the radial and tangential forces acting on the bearing when t = 1s.

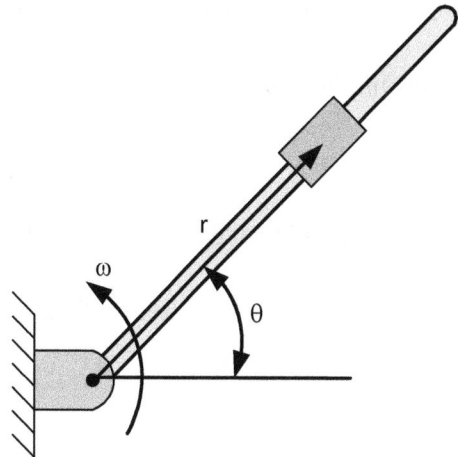

Fig. E5.8 Cylindrical bar rotating in the horizontal plane without friction.

Solution:

Prepare a free body diagram of the linear bearing as shown in Fig. E5.8a.

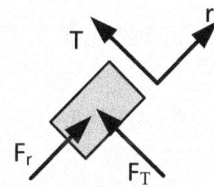

Fig. E5.8a A free body diagram of the linear bearing.

Write the equation for the forces in the radial and transverse directions.

ΣF_θ : $F_T = ma_T$ (a)

ΣF_r : $F_r = ma_r$ (b)

Note that the weight of the linear bearing is not included in the free body diagram, because the rod is rotating in the horizontal plane and the bearing slides without friction. Even in the absence of friction, a radial force is required to accelerate the linear bearing along the shaft in the radial direction.

The components of acceleration are given by:

$$a_r = [d^2r/dt^2 - r(d\theta/dt)^2]$$ (c)

$$a_T = [r\, d^2\theta/dt^2 + 2(d\theta/dt)(dr/dt)]$$ (d)

Recall $r = 3t^2 - t^3$ then $dr/dt = 6t - 3t^2$ and $d^2r/dt^2 = 6 - 6t$ (e)

Recall $\theta = 2t^2$ then $d\theta/dt = 4t$ and $d^2\theta/dt^2 = 4$ (f)

At t = 1 s we evaluate r, dr/dt, d^2r/dt^2, θ, $d\theta/dt$ and $d^2\theta/dt^2$ as:

$r = 3(1) - 1 = 2$ ft $dr/dt = 6(1) - 3(1) = 3$ ft/s $d^2r/dt^2 = 6 - 6(1) = 0$ ft/s^2 (g)

$\theta = 2t^2 = 2(1) = 2$ rad $\qquad$ $d\theta/dt = 4(1) = 4$ rad/s $\qquad$ $d^2\theta/dt^2 = 4$ rad/s^2 $\qquad$ (h)

Substituting theses values into Eqs. (c) and (d) gives:

$$a_r = [d^2r/dt^2 - r(d\theta/dt)^2] = 0 - 2(4)^2 = -32 \text{ ft/s}^2 \qquad (h)$$

$$a_T = [r\, d^2\theta/dt^2 + 2(d\theta/dt)(dr/dt)] = 2(4) + 2(4)(3) = 32 \text{ ft/s}^2 \qquad (i)$$

Then from Eqs (a) and (b), we find:

$$F_T = ma_T = (5/32.17)(32) = 4.974 \text{ lb} \qquad (j)$$

$$F_r = ma_r = (5/32.17)(-32) = -4.974 \text{ lb} \qquad (k)$$

The negative sign for F_r indicates that the force is directed toward the pin about which the rod is rotating.

EXAMPLE 5.9

A cylinder with a mass of 5 kg is moved along a horizontal channel by a rod that rotates with a constant angular velocity $\omega = 2$ rad/s as shown in Fig. E5.9. The angle θ of the rod with the vertical axis at this instant in time is 30°. The rotation occurs in the vertical plane. Determine the force of the rod acting on the cylinder and the normal force between the cylinder and the channel.

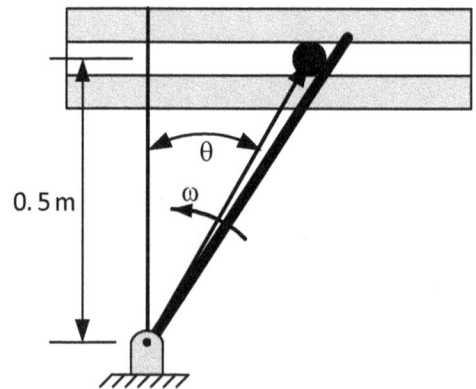

Fig. E5.9 A rod moving a cylinder with a mass of 5 kg along a horizontal channel.

Solution:

Prepare a free body diagram of the cylinder as shown in Fig. E5.9a.

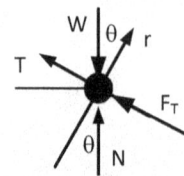

Fig. E5.9a A free body diagram of the cylinder.

Write the equation for the forces in the radial and transverse directions.

$$\Sigma F_r : \qquad\qquad - N \cos \theta + W \cos \theta = ma_r \qquad\qquad (a)$$

$$\Sigma F_T : \qquad\qquad F_T - W \sin \theta + N \sin \theta = ma_T \qquad\qquad (b)$$

Next use the geometry of the mechanism to determine the acceleration component a_θ.

$$r \cos \theta = 0.5$$
$$r = 0.5 \sec \theta$$

At $\theta = 30^O$ $\qquad\qquad r = 0.5 \sec 30^O = 0.5774 \text{ m} \qquad\qquad (c)$

$$dr/dt = 0.5(\sec \theta \tan \theta)(d\theta/dt) = 0.5(2)(\sec \theta \tan \theta) = 1.0 \,(\sec \theta \tan \theta)$$

At $\theta = 30^O$ $\quad dr/dt = 1.0 \,(\sec \theta \tan \theta) = 1.0 \,(1.155)(0.5774) = 0.6667 \text{ m/s} \qquad (d)$

$$d^2r/dt^2 = 1.0 \,[\sec \theta \sec^2 \theta \,(d\theta/dt) + \sec \theta \,(\tan^2 \theta)(d\theta/dt)]$$

$$d^2r/dt^2 = 1.0 \,(2)[\,\sec \theta \sec^2 \theta + \sec \theta \,(\tan^2 \theta)] = 2 \sec \theta[\,\tan^2 \theta + \sec^2 \theta]$$

At $\theta = 30^O$ $\quad d^2r/dt^2 = 2\,(1.154)[0.3333 + (1.154)^2] = 2.309 \,[\,1.667] = 3.848 \text{ m/s}^2 \qquad (e)$

Substituting results for r, dr/dt and d^2r/dt^2 into Eq. (5.11) yields:

$$a_r = [d^2r/dt^2 - r(d\theta/dt)^2] = 3.8484 - 0.5774(2)^2 = 1.539 \text{ m/s}^2 \qquad\qquad (f)$$

Substituting results for r, dr/dt and d^2r/dt^2 into Eq. (5.12) yields:

$$a_T = [r \,d^2\theta/dt^2 + 2(d\theta/dt)(dr/dt)] = 0 + 2(2)(0.6667) = 2.667 \text{ m/s}^2 \qquad (g)$$

Substitute numerical values into Eq. (a) to obtain:

$$- N \cos 30^O + 5\,(9.807) \cos 30^O = 5(1.539)$$

$$N = 40.15 \text{ N} \qquad\qquad (h)$$

Substitute numerical values into Eq. (b) to obtain:

$$F_T = m \,a_T + W \sin \theta - N \sin \theta$$

$$F_T = 5(2.667) + 5(9.807)(0.5) - 40.15(0.5)$$

$$F_T = 13.34 + 24.52 - 20.07 = 17.79 \text{ N} \qquad\qquad (i)$$

5.5   SUMMARY

In this chapter, we have considered a particle subjected to one or more forces. With a particle, we do not have to consider moments and rotation, because the motion involves only translation. The translation of a particle is controlled by Newton's second law that was introduced in Chapter 1.

$$\Sigma \mathbf{F} = m\frac{d\mathbf{v}}{dt} = m\mathbf{a} \tag{5.1}$$

The motion is referenced to an inertial coordinate system Oxyz, which is fixed in space.

Newton's second and third laws and his gravitational law are:

$$F = m\,g = W \tag{5.2}$$

and

$$g = GM\,/R_e{}^2 \tag{5.3}$$

If the particle is Earth bound, g is the gravitational constant, W is the weight of the body and g is related to the universal gravitational constant G, the mass of the earth (M), and the radius of the Earth R_e.

In the International System of Units (SI), the mass is given in kilograms (kg) and the unit for force is a newton (N).

In the U. S. Customary System, the mass is expressed in slugs and the unit for force is a pound (lb.)

We considered three different coordinates systems for dealing with examples for particle motion subjected to applied forces. For the rectangular coordinates, we write the equation of motion in a vector format as:

$$\Sigma F_x\,\mathbf{i} + \Sigma F_y\,\mathbf{j} + \Sigma F_z\,\mathbf{k} = m(a_x\,\mathbf{i} + a_y\,\mathbf{j} + a_z\,\mathbf{k}) \tag{5.4}$$

The scalar versions of the equations of motion in rectangular coordinates are:

$$\Sigma F_x = ma_x \qquad \Sigma F_y = ma_y \qquad \Sigma F_z = ma_z \tag{5.5}$$

If a particle is moving along a curvilinear path in the x-y plane, we often employ normal and tangential components in writing the equation of motion as:

$$\Sigma F_t\,\mathbf{u_t} + \Sigma F_n\,\mathbf{u_n} = ma_t + ma_n \tag{5.6}$$

where $\mathbf{u_t}$ and $\mathbf{u_n}$ are unit vectors in the tangential and normal directions, respectively and the tangential a_t and normal a_n components of acceleration are:

$$a_t = dv/dt \tag{5.7}$$

$$a_n = v^2/\rho \tag{5.8}$$

where ρ is the radius of curvature for the curvilinear path and the direction of the normal component of acceleration is toward the origin for ρ.

The scalar equations of motion in tangential and normal components of acceleration are:

$$\Sigma F_t = ma_t \qquad\qquad \Sigma F_n = ma_n \qquad\qquad (5.9)$$

When a particle is moving along a curvilinear path in space, we often employ cylindrical coordinates and write the equation of motion as:

$$\Sigma F_r \, \mathbf{u_r} + \Sigma F_T \, \mathbf{u_T} + \Sigma F_z \, \mathbf{k} = ma_r + ma_T + ma_z \qquad\qquad (5.10)$$

where $\mathbf{u_r}$ and $\mathbf{u_T}$ and $\mathbf{k}$ are unit vectors in the radial, transverse and vertical directions respectively.

We write the components of acceleration in the radial and transverse directions as:

$$a_r = [d^2r/dt^2 - r(d\theta/dt)^2] \qquad\qquad (5.11)$$

$$a_T = [r \, d^2\theta/dt^2 + 2(d\theta/dt)(dr/dt)] \qquad\qquad (5.12)$$

The scalar equations of motion in cylindrical coordinates are:

$$\Sigma F_r = ma_r \qquad\qquad \Sigma F_T = ma_T \qquad\qquad \Sigma F_z = ma_z \qquad\qquad (5.13)$$

When solving problems involving any of the coordinate systems, we draw a free body diagram of the particle after selecting the appropriate coordinate system. We then write equations for ΣF based on this diagram. Then we compute the magnitude of the acceleration components based on knowledge of the curvilinear path of the particle, using the appropriate equations that express the accelerations in terms of the linear and angular velocities. Combining the ΣF equations with the components of acceleration yields two or three relations that are solved to determine the unknown forces acting on the particle.

CHAPTER 6

KINETICS OF RIGID BODY MOTION

INTRODUCTION

In this chapter you will study rigid body motion that includes both translation and rotation. You will find that problems involving translation are solved using Newton's second law. Problems involving rotation are solved using a similar equation that relates the sum of the moments to the angular acceleration. You will need to solve for the mass moment of inertia when dealing with rotation. In many cases, you will find that the parallel axis theorem is very useful if you have information on the mass moment of inertia about the centroidal axes of the rigid body.

When dealing with the translation of rigid bodies, you determine the center of gravity of the body and refer accelerations to that point. If the rigid body is translating without rotation, you know that the sum of the moments about the center of gravity of the body vanishes. In rectilinear motion, the acceleration components are referenced to the x and y axes. With curvilinear motion, but without rotation, the acceleration components are referenced to the normal and tangential components.

When you treat centroidal rotation of a plane rigid body subjected to a system of forces, recognize that these forces create a moment about the centroid. The moment vector in turn produces an angular acceleration α of the body that is easily determined if the mass moment of inertia about the centroid is known. Because the body is rotating, the tangential and normal components of the acceleration are determined from well know relations $a_t = r\,\alpha$ and $a_n = r\,\omega^2$.

With non-centroidal rotation of a plane body the center of mass moves along a circular path and the normal and tangential components of acceleration depend on the radius of that circular path. In the general case of plane motion of rigid bodies, you observe the body rotating and translating. In this case, the equations of motioned are referenced to an x-y axes and the moments are taken about the centroid of the body.

CONCEPT PROBLEM

Let's consider a thin disk with a mass of 14 kg and a 750 mm radius that is supported by a cable wrapped around its circumference, as shown below. When the disk is released from rest, it rotates in the clockwise direction and accelerates downward. You are to determine the linear and angular acceleration of the disk and the force on the cable, as the disk moves downward.

A disk initially at rest rotates and translates as it moves downward.

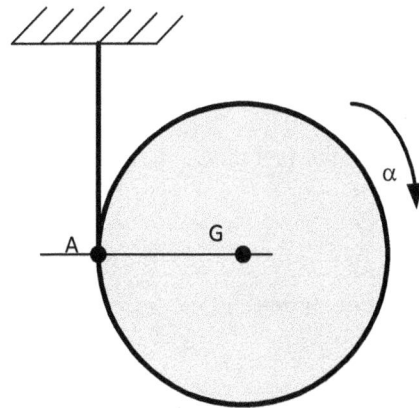

You begin by preparing a free body diagram of the disk. You should include the inertia terms for both rotation and translation, as shown below: When adding the inertia force and inertia moment to the free body diagram, you reversed their direction and noted this change by using dashed lines in their representation. With the applied forces, inertia force and inertia moment added to the free body diagram, you have created a state of equilibrium for the disk, where the sum of the forces and moments equal zero.

The free body diagram of the disk showing the applied forces F due to the cable, the weight W as well as the inertia force ma_G and the inertia moment $I_G \alpha$.

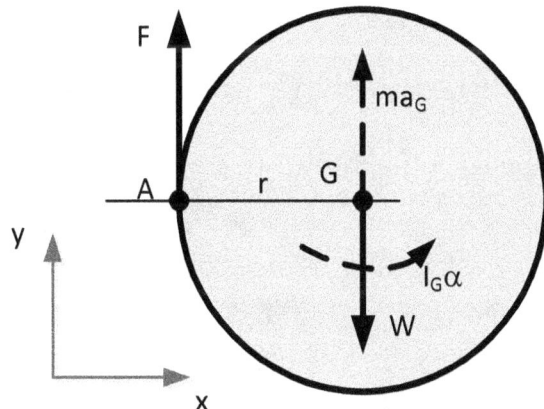

You begin by writing the equilibrium equation for the moments about point A:

$$\Sigma M_A = 0 \qquad (a)$$

$$\Sigma M_A: \qquad m\, a_G\, r - m\, g\, r + I_G\, \alpha = 0 \qquad (b)$$

where

$$I_G = \tfrac{1}{2}\, mr^2 \qquad (c)$$

and

$$a_G = r\, \alpha \qquad (d)$$

You substitute Eqs. (c) and (d) into Eq. (a) and simplify to obtain:

$$r\, \alpha - g + \tfrac{1}{2}\, r\, \alpha = 0 \qquad (e)$$

Substitute numerical values into Eq. (e) and solve for α to obtain:

$$\alpha = (2/3)\ g/r = (2/3)(9.807/0.75) = 8.717 \text{ rad/s}^2 \qquad \text{(f)}$$

Next use the fact that $\Sigma M_G = 0$, which leads to:

$$r\ F = I_G\ \alpha \qquad \text{(g)}$$

Substitute numerical values into Eq. (g) and solve for F to obtain:

$$0.75\ F = \tfrac{1}{2}\ mr^2\ \alpha = \tfrac{1}{2}(14)(0.75)^2\ (8.717)\ 34.32 \text{ N}$$

$$F = 7(0.75)\ (8.717) = 45.76 \text{ N} \qquad \text{(h)}$$

You have solved a difficult problem involving a disc subjected to both rotation and translation motion. Your approach was consistent. You prepared a free body diagram and made certain to include all of the forces involved. You decided to include the inertia force and the inertia moment in your free body diagram. This choice enabled you to use the equilibrium equations. There were several ways to write the equilibrium relations and you selected to use $\Sigma M_A = 0$ and $\Sigma M_G = 0$. The first of these equations enabled you to solve for the angular acceleration α and the second enabled you to solve for the tension F in the cable. You can also solve for the force F by using $\Sigma F_y = 0$, which leads to:

$$\Sigma F_y: \qquad\qquad F + ma_G - W = 0 \qquad \text{(i)}$$

From Eqs. (d) and (f) we write:

$$a_G = r\ \alpha = r\ (2/3)(g/r) = (2/3)g \qquad \text{(j)}$$

Substituting Eq. (j) into Eq. (i) and noting $W = mg$ yields:

$$F = m\ [g - (2/3)g] = (14/3)g = (14/3)9.807 = 45.77 \text{ N} \qquad \text{(k)}$$

In both approaches the correct answer can be obtained by carefully noting the direction of the forces and/or the moments and properly executing the solution.

Before proceeding let's consider three types of motion:

Translation:

If a rigid body is in planar translation, the angular acceleration of the body is zero and the external forces acting on it reduce to the vector $m\mathbf{a_G}$ applied to the body at the center of gravity G as shown below:

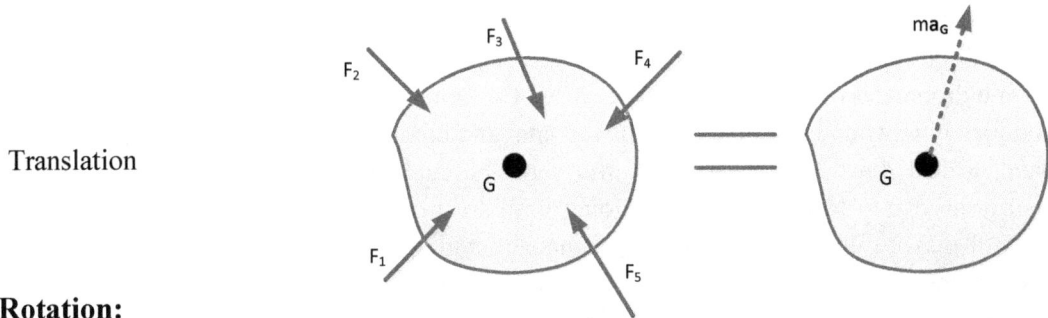

Translation

Centroidal Rotation:

Centroidal rotation occurs when a body, which is symmetrical relative to the reference plane OXY, rotates about a fixed axis passing through the body and in the Z direction. Because the acceleration $\mathbf{a_G} = 0$, the effect of all the forces acting on the body reduce to the moment $\mathbf{M_G} = I_G\,\boldsymbol{\alpha}$ as shown below:

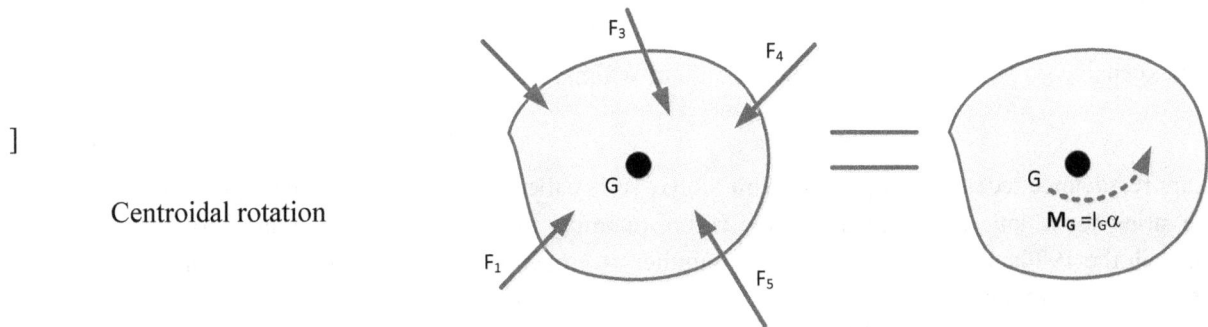

]

Centroidal rotation

General Plane Motion:

The case of general plane motion of a rigid body, which is symmetrical with respect to the reference plane OXY, can be replaced by combining the motion due to translation and centroidal rotation, as shown below: We note that the mass center G of the rigid body moves as if the entire mass of the body was concentrated at G and as if the resultant force acted at this point. The motion of the body is controlled by both the resultant force and the resultant moment of the external forces about the center of gravity G. Hence, a rigid body in plane motion rotates about its mass center or its instantaneous center, while it is undergoing translation.

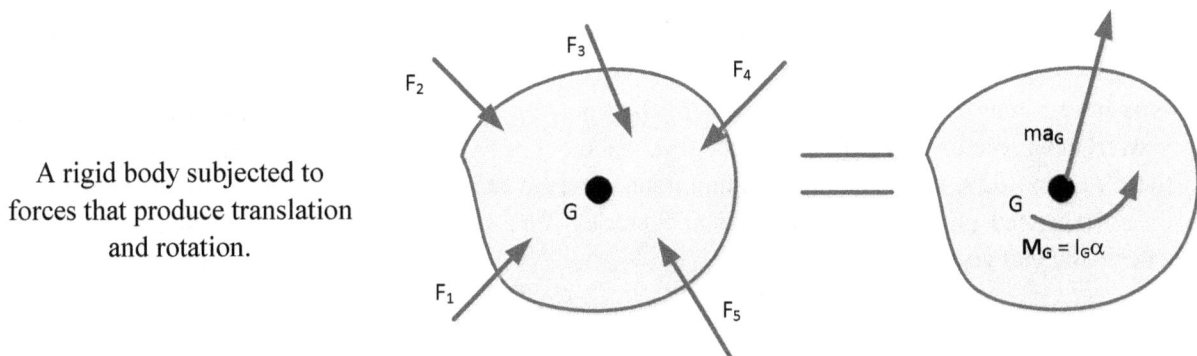

A rigid body subjected to forces that produce translation and rotation.

DISCUSSION

Back in the late 1930s, most of the kids in my neighborhood owned a Yo-Yo. It was a simple toy consisting of a disk about 2.5 in. in diameter and about 1.0 in thick. A deep grove was cut into the disk leaving a small diameter shaft at its center. A string about 2 to 2.5 ft long was attached to this shaft. This string had a loop at the other end, into which we inserted a couple of fingers. We wound the string about the shaft until the disc was in our hand. We release the disk giving it a modest velocity downward. The Yo-Yo dropped downward spinning, because of the moment generated by the string. When the string was completely unwound, the disc continued to spin and climb the string back into our hand. By moving our hand up and down with each cycle, we were able add the energy necessary to keep the yo-yo going until we became tired of playing with it. This simple method of cycling the disc up and down is called looping. A photograph of a few inexpensive yo-yos is presented to the right:

Inexpensive yo-yo's available for about $5.00 at Walmart.

A more advance technique of playing with the yo-yo is called sleeping, where the yo-yo spins at the end of the string for a noticeable amount of time before returning to the hand. As popularity of yo-yos spread through the 1970s and 1980s, there were a number of innovations in the technology involving the ability of the axle to rotate relative to the disc. A rotating axle require several changes from the simple wooden discs and axle. The new designs incorporated metal axles that were supported by small roller bearings. A photograph of an all metal yo-yo with roller bearings is shown below:

An all metal Yo-Yo available on Amazon.com.

In yo-yos with bearings, the friction is reduced significantly, while the yo-yo is spinning. This reduction in frictional losses enables the player to perform longer and more complex tricks with the all metal toys.

Yo-yos have a long history with evidence of a boy playing with one in Greece in 440 BC. Historical records from Greece for that period describe toys made out of wood, metal or fired clay. The yo-yo came to the U. S. in 1929, when a Filipino immigrant started a manufacturing company in California. The business expanded rapidly so two additional factories were started which employed 600 workers and produced 300,000 yo-yos daily.

6.1   INTRODUCTION

This chapter deals with rigid body motion including both translation and rotation. In Chapter 4, we showed methods for dealing with rigid body translation; however, to analyze both translation and rotation we employ two equations of motion, which are:

$$\Sigma \mathbf{F} = m\mathbf{a} \qquad\qquad \text{for translation} \qquad\qquad (6.1)$$

$$\Sigma \mathbf{M} = I\boldsymbol{\alpha} \qquad\qquad \text{for rotation} \qquad\qquad (6.2)$$

where **M** is the moment created by the application of the forces and **α** is the angular acceleration of the rigid body. I is the mass moment of inertia or the second moment of the mass of the rigid body.

When applying the equations of motion it is essential that the acceleration of the body be referenced to a coordinate system that is either fixed in space or moving with a constant velocity. It is common practice to fix the reference coordinates to the stars and to refer to this coordinate system as an inertia system of coordinates.

6.2   MASS MOMENT OF INERTIA

The mass moment of inertia I is the second moment of all the mass elements about a specified axis. We write the equation for I_z in terms of the mass of the rigid body as:

$$I_z = \int_m r^2 dm = \int_m (x^2 + y^2) dm \qquad\qquad (6.3a)$$

$$I_x = \int_m r^2 dm = \int_m (y^2 + z^2) dm \qquad\qquad (6.3b)$$

$$I_y = \int_m r^2 dm = \int_m (x^2 + z^2) dm \qquad\qquad (6.3c)$$

where r is the radial distance from the axis to the element dm.

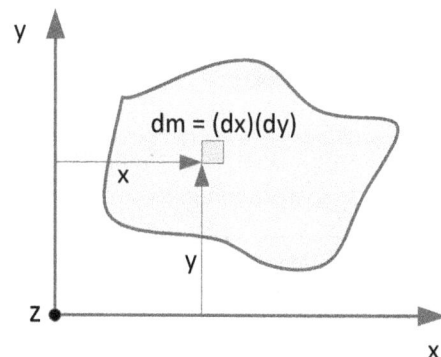

Fig. 6.1 Graphical representation of an element involved in the calculation of the mass moment of inertia of a plane body about the z axis.

We can also write the equation for I_z in terms of a volume element dV as:

$$I_z = \int_V \rho r^2 dV \qquad (6.4)$$

where ρ is the mass density of the material used in fabricating the rigid body, $dm = \rho dV$ and $r^2 = x^2 + y^2$.

We sometimes use the radius of gyration k which is related to the second moment of inertia as:

$$I = mk^2 \qquad (6.4a)$$

For a three dimensional body, we use Eqs. (6.3a to c) to compute the mass moment of inertia, but reference a three dimensional coordinate system, as shown in Fig. 6.2.

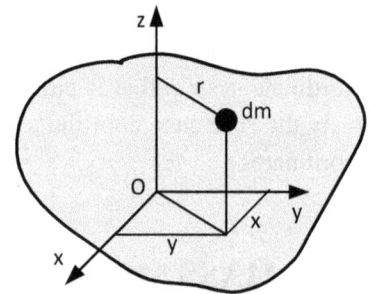

Fig. 6.2 A Oxyz coordinate system on a rigid body, where the z coordinate is normal to the plane of the rigid body.

Let's consider two examples to demonstrate the use of integral calculus to determine the mass moment of inertia.

EXAMPLE 6.1

Determine the mass moment of inertia with respect to the x, y and z axes for the thin rectangular plate, as shown in Fig. E6.1.

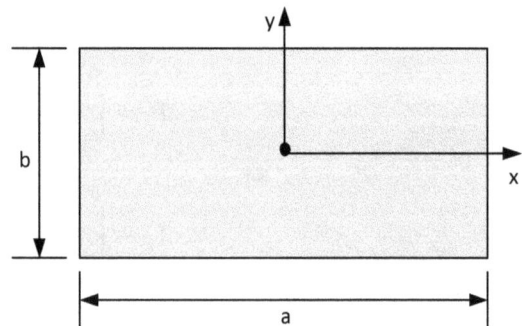

Fig. E6.1 A thin rectangular plate. The z axis is normal to the plate.

Solution:

To determine the mass moment of inertia relative to the x axis, prepare a drawing of the plate and construct the elemental area dA and note that $r = y$ when calculating I_x.

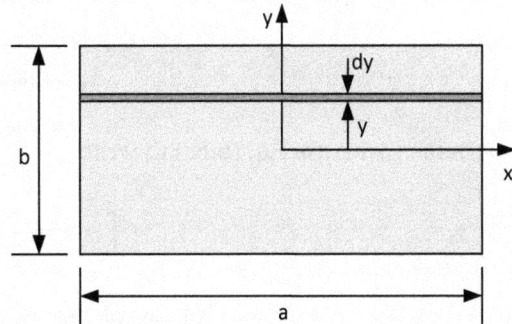

Fig. E6.1a A drawing of the plate showing the
elemental area dA = ady.

Write the relation for I_x as:

$$I_x = \rho t \int_A r^2 dA$$

where ρ is the mass density of the plate's material and t is the thickness of the plate and dm = ρt dA.

$$I_x = 2\rho t \int_0^{b/2} y^2 a \; dy = 2\rho t \; a \left[\frac{y^3}{3} \right]_0^{b/2} \qquad (a)$$

In Eq. (a) we made use of symmetry and integrated from 0 to b/2. Substituting limits into Eq. (a) yields:

$$I_x = \rho t a b^3 / 12 \qquad (b)$$

Note that ρtab = m the mass of the plate.

Then
$$I_x = m b^2 / 12 \qquad (c)$$

To determine the mass moment of inertia about the y axis, we repeat the process by preparing a drawing of the plate that shows the elemental area bdx.

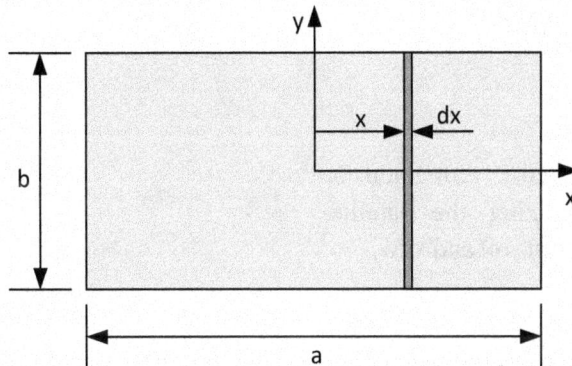

Fig. E6.1b A drawing of the plate showing the
elemental area dA = b dx.

Then it is evident that:

$$I_y = 2\rho t \int_0^{a/2} x^2 b \, dy = 2\rho t \; b \left[\frac{x^3}{3} \right]_0^{a/2} \qquad (d)$$

In Eq. (d) we made use of symmetry and integrated from 0 to a/2. Substituting limits into Eq. (d) yields:

$$I_y = \rho t a^3 b / 12 \qquad \text{(e)}$$

or

$$I_y = m a^2 / 12 \qquad \text{(f)}$$

To determine I_{zz} we use Eq. (6.6a) to write:

$$I_z = \int_m r^2 dm = \int_m (x^2 + y^2) dm \qquad \text{(g)}$$

and

$$I_z = I_y + I_x = (\rho\, t\, a\, b / 12)(a^2 + b^2) = (m/12)(a^2 + b^2) \qquad \text{(h)}$$

EXAMPLE 6.2

Significant changes have occurred in recent years regarding satellites in space. New rockets have been developed by Space X and Blue Origins that are lower in cost, when compared to the legacy rockets used in the past. In addition electronic components are smaller in size, weigh less and have much more capability. These developments have led to the use of much smaller satellites today, when compared to the large and heavy satellites that were common in the past. This preamble leads us to the problem statement. Sketch the design of a system to stabilize a satellite in space using the minimum weight in your design.

Solution:

The system will employ folding arms so that the hardware fits into the launch package without adding significant volume. When the satellite is placed on station, the folding arms deploy and the two masses that serve to stabilize the satellite are at a distance L from the center of mass of the satellite, as shown in Fig. E6.2.

Fig. E6.2 A method for stabilizing the satellite against roll and yaw.

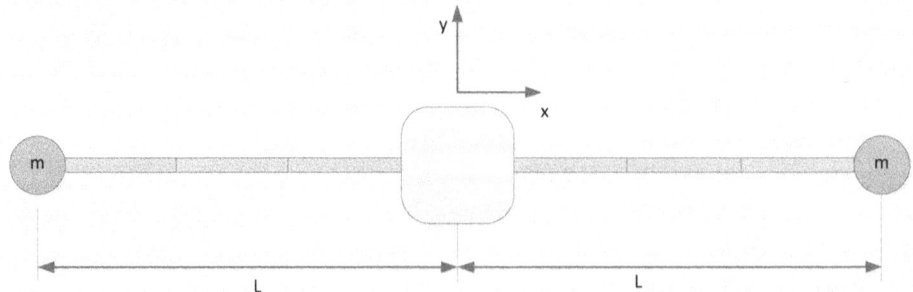

The mass moment of inertia due to two masses located at the ends of the folding arms is $I_y = I_z = 2mL^2$. Because L is large and L^2 is much larger, the increase in I_y and I_z is significant and the satellite is stabilized against both roll and yaw. But what about pitch. To stabilize the satellite against pitch, we add two more masses with folding arms, as shown in Fig. E6.2a.

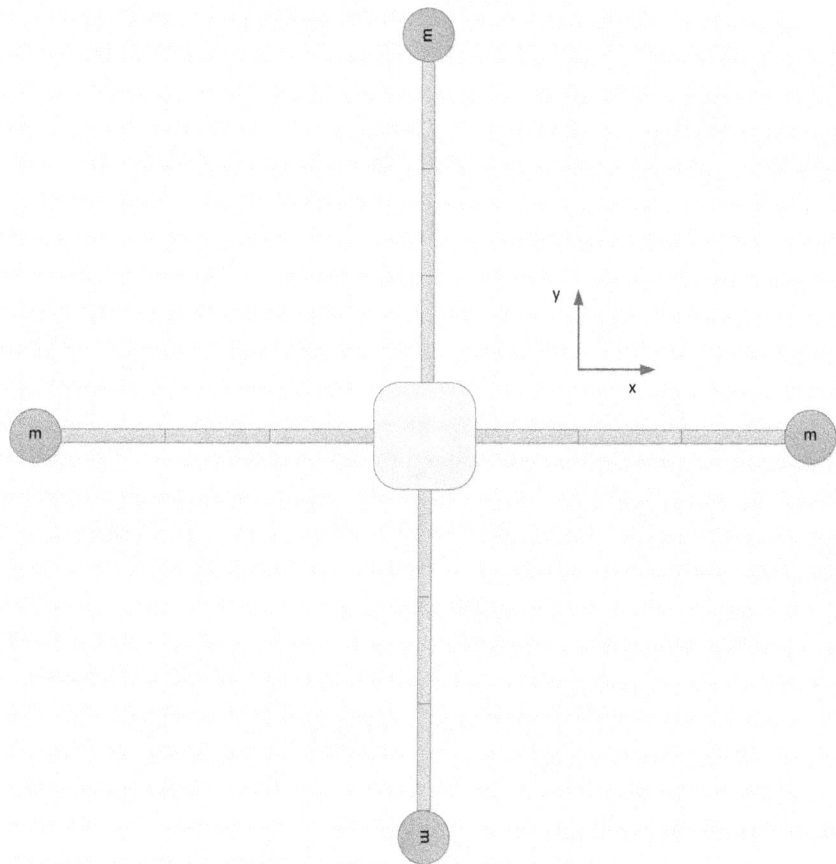

Fig. 6.2a Arms and masses
were added to stabilize the
satellite for pitch oscillations.

The masses at the ends of the arms are filled with payload (usually electronic systems and fuel. The fuel is used over the lifetime of the satellite to power small rocket engines that are fired to control the attitude and position of the satellite in space.

EXAMPLE 6.3

Derive the equations for I_x, I_y and I_z for a thin circular disk, with a thickness t as, shown in Fig. E6.3. The z axis is located at the center of the disc and is directed out of the paper.

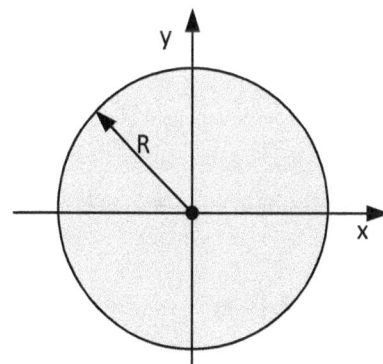

Fig. E6.3 A thin circular disc with a mass density ρ.

Solution:

Let's first determine I_z by using integration with polar coordinates. The element $dA = r \, dr \, d\theta$, as shown in Fig. E6.3a.

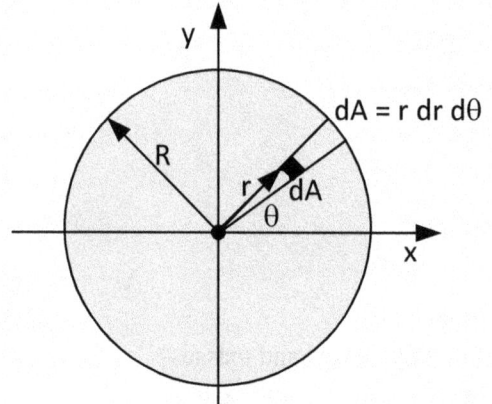

Fig. E6.3a The elemental area $dA = r \, dr \, d\theta$ as shown.

We use Eq. (6.6a) for I_z and write:

$$I_z = \rho t \int_A r^2 dA \qquad \text{(a)}$$

Substituting $dA = r \, dr \, d\theta$ into Eq. (a) gives:

$$I_z = \rho t \int_0^{2\pi} d\theta \int_0^R r^3 dr \qquad \text{(b)}$$

Integrate and simplify to obtain:

$$I_z = \frac{\rho t \pi R^4}{2} \qquad \text{(c)}$$

Substitute $m = \rho \, t \, \pi \, R^2$ into Eq. (c) to obtain:

$$I_z = m R^2/2 \qquad \text{(d)}$$

Recall Eq. (6.6) and note that:

$$I_z = I_x + I_y \qquad \text{(e)}$$

Due to symmetry:

$$I_x = I_y \qquad \text{(f)}$$

Then from Eqs. (d) and (f):

$$I_x = I_y = mR^2/4 \qquad \text{(g)}$$

6.3   PARALLEL AXIS THEORM

The parallel axis theorem for the mass moment of inertia is similar to the parallel axis theorem for the area moment of inertia. To derive this useful equation, consider the body shown in Fig. 6.3, where two parallel coordinate systems are drawn. The first system $Gx_Gy_Gz_G$ is a centroidal system with G at the center of mass of the rigid body. The second coordinate system Oxyz is position some arbitrary distance from the centroidal system. Point A, located within the rigid body, is drawn on the x-y plane of the Oxyz coordinate system. We have also identified the distances between point A and the origin G of the centroidal system with dimensions $\overline{x}$, $\overline{y}$ and $\overline{z}$.

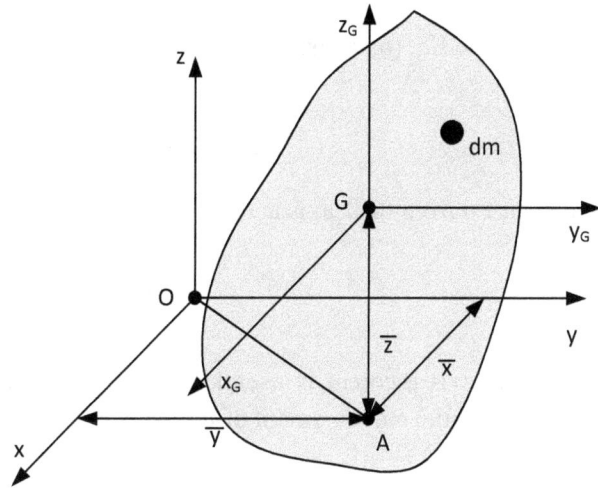

Fig. 6.3 A rigid body located relative to two
parallel coordinate systems.

The elemental mass dm is located relative to the Oxyz coordinate system by:

$$x = x_G + \overline{x} \qquad\qquad y = y_G + \overline{y} \qquad\qquad z = z_G + \overline{z} \qquad\qquad (a)$$

Next let's write the equation for I_z relative to the Oxyz coordinate system as:

$$I_z = \int_m (x^2 + y^2)dm = \int_m [(x_G + \overline{x})^2 + (y_G + \overline{y})^2]dm \qquad\qquad (b)$$

Expand Eq. (b) to obtain:

$$I_z = \int_m \left[x_G^2 + 2x_G\overline{x} + \overline{x}^2 + y_G^2 + 2y_G\overline{y} + \overline{y}^2 \right]dm \qquad\qquad (c)$$

Rearranging terms in Eq. (c) gives:

$$I_z = \int_m [x_G^2 + y_G^2]dm + 2\overline{x}\int_m x_G dm + 2\overline{y}\int_m y_G dm + (\overline{x}^2 + \overline{y}^2)\int_m dm \qquad\qquad (d)$$

Equation (d) can be rewritten as:

$$I_z = \overline{I}_z + m(\overline{x}^2 + \overline{y}^2) \qquad\qquad (6.5)$$

where $\overline{I}_z = \int_m [x_G^2 + y_G^2]dm$ is the mass moment of inertia about the centroidal z axis.

Note that the terms $\int_m x_G dm$ and $\int_m y_G dm$ both vanish, because they are first moments of the mass about the centroidal axes. It is evident then that we may write:

$$I_x = \overline{I}_x + m(\overline{z}^2 + \overline{y}^2) \tag{6.6}$$

$$I_y = \overline{I}_y + m(\overline{x}^2 + \overline{z}^2) \tag{6.7}$$

We can rewrite Eq. (6.7a) as:

$$I_z = \overline{I}_z + m\,d^2 \tag{6.8}$$

The distance d from the z axis and the z_G axis is given by:

$$d = \sqrt{(\overline{x}^2 + \overline{y}^2)} \tag{6.9}$$

The parallel axis theorem is useful in determining the mass moment of inertia of a body about some parallel axis, if the mass moment of inertia is known for the body referenced to its centroidal axes.

6.4   MASS MOMENT OF INERTIA: THREE DIMENSIONAL BODIES

We will demonstrate the method of determining the mass moment of inertia, for three-dimensional rigid bodies by providing the two examples shown below:

EXAMPLE 6.4

Determine the mass moment of inertia of a long slender cylindrical rod of length L, with respect to a perpendicular axis located at the end of the rod.

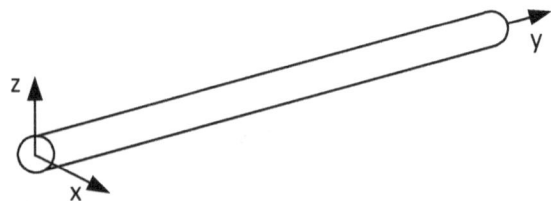

Fig. E6.4 A long slender rod of length L.

Solution:

Prepare a drawing of the element dm and add dimensions locating this element on the long slender rod.

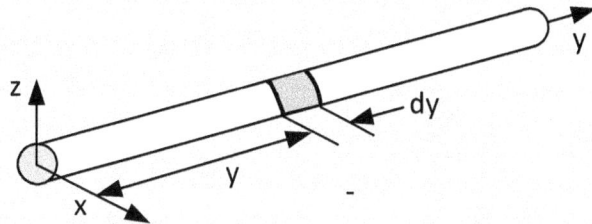

Fig. E6.4a An elemental length dy.

If the entire rod has a mass of m and a length of L, we can write dm = (m/L)dy.

Then we use Eq. (6.3) and write:

$$I_x = \int_m y^2 dm = \frac{m}{L}\int_0^L y^2 dy \qquad\qquad (a)$$

Integrate and simplify to obtain:

$$I_x = mL^2/3 \qquad\qquad (6.10)$$

EXAMPLE 6.5

Determine the mass moment of inertia for the rectangular bar, shown in Fig. E6.5, relative to the x axis. The bar's length is L, its height is H and its width is W.

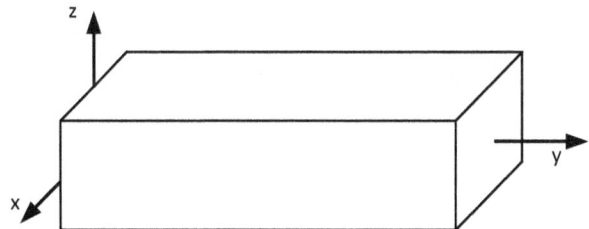

Fig. E6.5 A rectangular bar with length L, width
W and height H.

Solution:

Prepare a drawing showing a slice of the bar that can be used to define dm.

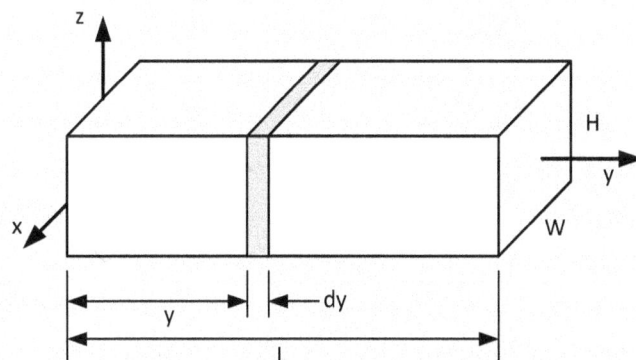

We may write an equation for the mass of the slice as:

$$dm = \rho \, H \, W \, dy \tag{a}$$

Recognizing the slice as a thin rectangular plate, we refer to Example 6.1, Eq. (6.5) and note the incremental mass moment of inertia about the centroidal axis at the position y is :

$$dI_{xG} = (H^2/12)dm \tag{b}$$

To determine the mass moment of inertia at $y = 0$, we employ the parallel axis theorem and write:

$$dI_x = dI_{xG} + y^2 \, dm \tag{c}$$

Substituting Eqs. (a) and (b) into Eq. (c) yields:

$$dI_x = \rho HW \, (H^2/12 + y^2)dy \tag{d}$$

Integrating Eq. (d) and substituting the limits from 0 to L for the range of y yields:

$$I_x = (\rho HWL/12) \, (H^2 + 4L^2) = (m/12)(H^2 + 4L^2) \tag{6.11}$$

Equation (6.11) gives the mass moment of inertia of a rectangular bar about a perpendicular axis located at one end of the bar.

It is evident that we can determine the mass moment of inertia I of three-dimensional bodies by integration. However, we can also refer to tables that show the results for the mass moment of inertia shapes commonly encountered. We show a list of these shapes in Table 6.1.

Table 6.1
The mass moment of inertial for select two and three dimensional shapes.

Thin Rectangular Plate

$$I_z = 1/12 \, m \, (a^2 + b^2)$$

$$I_x = 1/12 \, m \, b^2$$

$$I_y = 1/12 \, m \, a^2$$

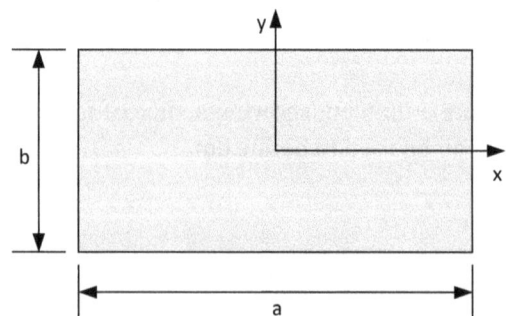

Thin Circular Disk

$I_z = 1/2\ m\ R^2$

$I_x = I_y = 1/4\ m\ R^2$

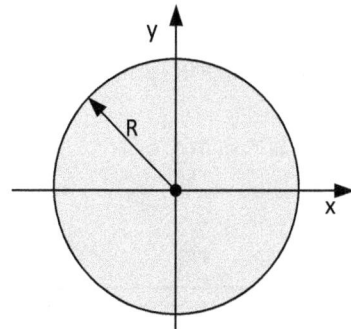

Thin Ring

$I_x = I_y = \tfrac{1}{2}\ m\ R^2$

$I_z = m\ R^2$

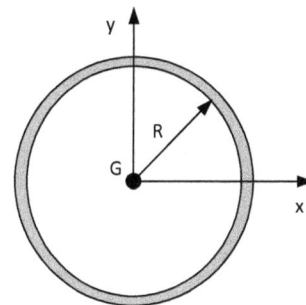

Long Slender Rod

$I_x = I_z = 1/3\ mL^2$ about its end

$I_x = I_z = 1/12\ mL^2$ about its centroid

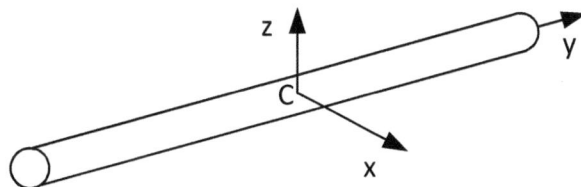

Rectangular Bar

$I_x = 1/12\ m(L^2 + H^2)$

$I_z = 1/12\ m(W^2 + L^2)$

$I_y = 1/12\ m(W^2 + H^2)$

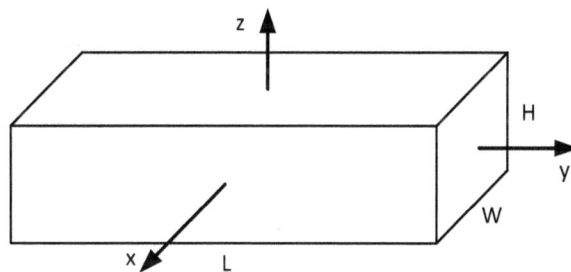

Circular Bar

$I_x = I_z = 1/12\ m(3R^2 + L^2)$

$I_y = \frac{1}{2}\ mR^2$

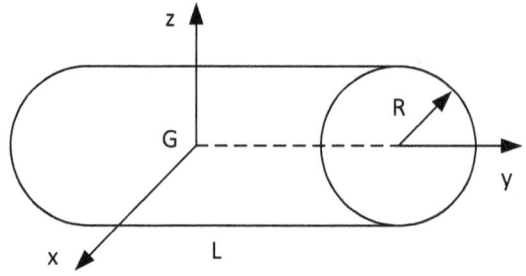

Circular Cone

$I_x = I_z = 3/5\ m(R^2/4 + L^2)$

$I_y = 3/10\ mR^2$

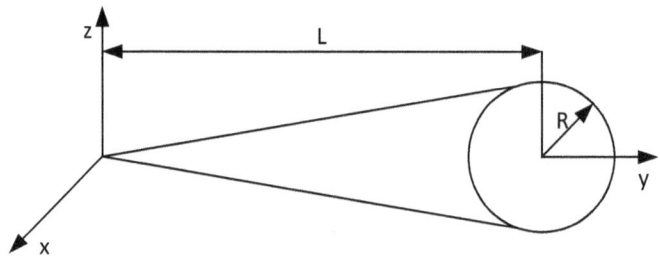

Sphere

$I_x = I_z = I_y = 2/5\ mR^2$

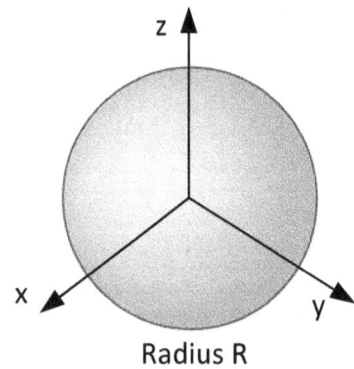

Hemisphere

$I_x = I_y = 0.259mR^2$

$I_z = 2/5\ mR^2$

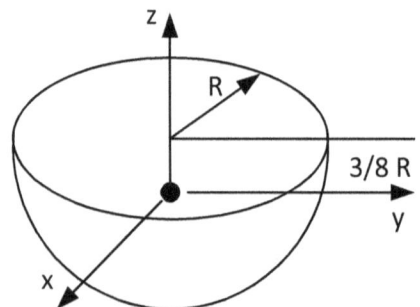

6.5   MASS MOMENT OF INERTIA: COMPOSITE BODIES

Many complex bodies can be divided into a number of different shapes that include those shown in Table 6.1. In this process, we can add or even subtract (for holes), if we take account of the axis to which the shapes are referenced and use the parallel axis theorem. Let's demonstrate this technique with two examples.

EXAMPLE 6.6

The rigid body, shown in Fig. E6.6, consists of a thin cylindrical rod 450 mm long, which is attached to a solid sphere with a radius of 100 mm. The cylindrical rod has a mass of 10 kg and the sphere has a mass of 15 kg. Determine the mass moment of inertial about an axis at point A that is directed out of the paper.

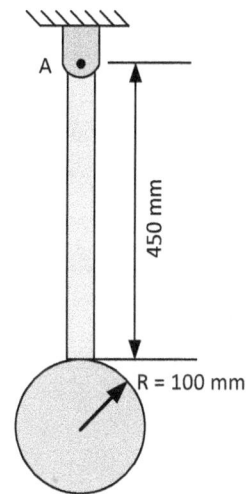

Fig. E6.6 A pendulum consisting of a thin cylindrical rod and a sphere.

Solution:

Begin by calculating the mass moment of inertia about the centroid of the sphere using formulas from Table 6.1.

$$I_s)_G = 2/5 \ m \ R^2 = (2/5)(15)(0.1)^2 = 0.060 \ \text{kg-m}^2 \qquad (a)$$

Next calculate the mass moment of inertia about the centroid of the cylindrical rod using formulas from Table 6.1.

$$I_R)_G = 1/12 \ m \ L^2 = (1/12)(10)(0.45)^2 = 0.1688 \ \text{kg-m}^2 \qquad (b)$$

Employ the parallel axis theorem to reference the mass moment of inertia to point A.

For the sphere $\qquad I_s)_A = I_s)_G + md^2 = 0.06 + 15(0.55)^2 = 4.598 \ \text{kg-m}^2 \qquad (c)$

For the rod $\qquad I_R)_A = I_R)_G + md^2 = 0.1688 + 10(0.225)^2 = 0.6751 \ \text{kg-m}^2 \qquad (d)$

Adding the results from Eqs. (c) and (d) yields:

$$I_A = I_s)_A + I_R)_A = 5.273 \ \text{kg-m}^2 \qquad (e)$$

EXAMPLE 6.7

A thin square plate with sides measuring 1.8 m and a thickness of 150 mm is supported at a corner, as shown in Fig. E6.7. The density of the plate material is 40 kg/m^3. A hole with a radius R = 250 mm is cut into the center of the plate. Determine the inertia of the plate relative to an axis normal to the plate and located at point A.

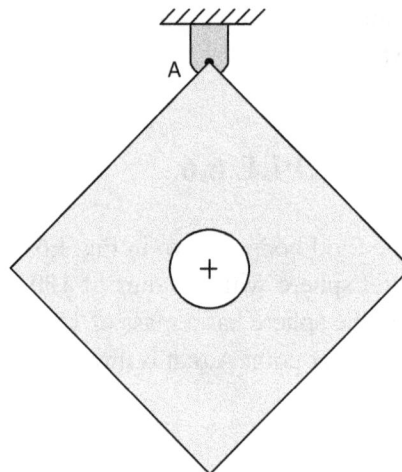

Fig. E6.7 A square plate with a thickness of 150 mm is supported at point A.

Solution:

Determine the mass of the plate and the mass of the material removed in cutting the hole.

The mass of the plate is:

$$m_p = A t \rho = (1.8)^2 (0.150)(40) = 19.44 \text{ kg} \tag{a}$$

The mass of the material removed for the hole is:

$$m_h = A t \rho = \pi R^2 t \rho = \pi(0.250)^2 (0.150)(40) = 1.178 \text{ kg} \tag{b}$$

Using the results from Table 6.1 we calculate the mass moment of inertia relative to the centroidal axis as:

$$I_p)_G = (1/12) m_p (a^2 + b^2) = (1/12)(19.44)(2)(1.8)^2 = 10.50 \text{ kg-m}^2 \tag{c}$$

$$I_h)_G = (1/2) m_h r^2 = (1/2)(1.178)(0.250)^2 = 0.0368 \text{ kg-m}^2 \tag{d}$$

Employ the parallel axis theorem to determine the mass moment of inertia for the two components at point A.

$$I_p)_A = I_p)_G + m_p d^2 = 10.50 + 19.44(1.8)(\sin 45^O) = 35.24 \text{ kg-m}^2 \tag{e}$$

$$I_h)_A = I_h)_G + m_h d^2 = 0.0368 + 1.178(1.8)(\sin 45^O) = 1.536 \text{ kg-m}^2 \tag{d}$$

Subtracting the contribution of the hole to the mass moment of inertia of the plate yields:

$$I_A = 35.24 - 1.536 = 33.70 \text{ kg-m}^2 \tag{f}$$

6.6   TRANSLATION OF RIGID BODIES

We have previously discussed translation of rigid bodies, but have not considered the distribution of the mass of the bodies. We return to the topic of translation in this section and rewrite Eq. (6.1) as:

$$\Sigma F = m\ a_G \qquad\qquad (6.12)$$

where the acceleration a refers to the center of mass of the rigid body.

From Eq. (6.12), we write the scalar versions of the translation equations of motion as:

$$\Sigma F_x = m\ a_G)_x \qquad\qquad (6.13a)$$

$$\Sigma F_y = m\ a_G)_y \qquad\qquad (6.13b)$$

Because the rigid body is translating without rotation, it is evident that:

$$\Sigma M_G = 0 \qquad\qquad (6.13c)$$

Equations (6.13) are employed when analyzing rigid bodies in rectilinear motion.

6.6.1 Curvilinear Motion

When a rigid body moves along a curved path, without rotation, we often resolve the acceleration vector into normal and tangential components. With curvilinear motion, without rotation, we express the scalar version of Eq. (6.12) as:

$$\Sigma F_n = m\ a_G)_n \qquad\qquad (6.14a)$$

$$\Sigma F_t = m\ a_G)_t \qquad\qquad (6.14b)$$

Because the rigid body is translating without rotation, it is evident that:

$$\Sigma M_G = 0 \qquad\qquad (6.14c)$$

Equations (6.14) are employed when analyzing rigid bodies in curvilinear motion without rotation.
To demonstrate the use of these scalar equations, we show solutions for four examples.

EXAMPLE 6.8

A large box measuring 6 ft high by 8 ft wide by 7 ft deep is moved along a level floor with an applied force F = 150 lb, as shown in Fig. E6.8. The box weighs 400 lb. The coefficient of friction is $\mu = 0.18$. Determine the acceleration of the box along the floor.

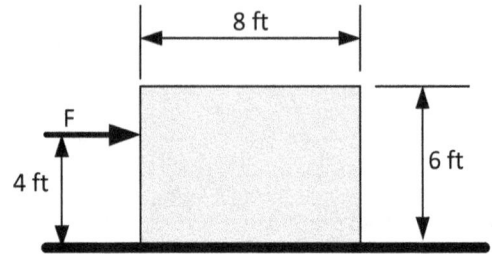

Fig. E6.8 A rectangular box is pushed along a level floor by a force F = 400 lb.

Solution:

Prepare a free body diagram of the box showing all the forces acting on it, as shown in Fig. E6.8a:

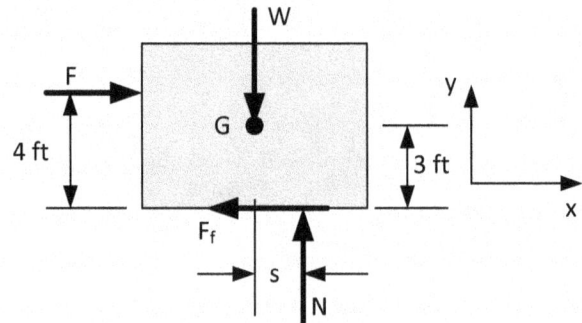

Fig. E6.8a The box with applied forces and a coordinate system.

Write the scalar equations of motion for the box:

$$\Sigma F_x = F - F_f = m\ a_G)_x = (400/32.17)\ a_G)_x = 12.43\ a_G)_x \qquad \text{(a)}$$

$$\Sigma F_y = N - W = 0 \qquad \text{and} \qquad N = W = 400\ \text{lb} \qquad \text{(b)}$$

$$\Sigma M_G = -1\ F + N\ s - 3F_f = 0 \qquad \text{(c)}$$

Substituting numbers into Eq. (c) and solving for the distance s yields.

$$s = (1\ F + 3\ F_f)/N = (1/400)[400 + 3(400)(0.18)] = 1.54\ \text{ft}. \qquad \text{(d)}$$

This result for s indicates that the box cannot tip and that Eqs. (a) and (b) are valid.

Substituting numerical values into Eq. (a) yields:

$$400 - 0.18(400) = 12.43\ a_G)_x \qquad a_G)_x = (328/12.43) = 26.39\ \text{ft./s}^2 \qquad \text{(e)}$$

EXAMPLE 6.9

A bar 20 ft. long with a square cross section with 2 ft. on its sides weighs 10,000 lb. The bar is lifted with a crane, as shown in Fig. E6.9. The acceleration of the bar during the lift is 0.5 ft/s². The lift is made with two symmetrical cables each located 5 ft from the ends of the bar and oriented 45° with axis of the bar. Determine the moment M_G at the center of the bar.

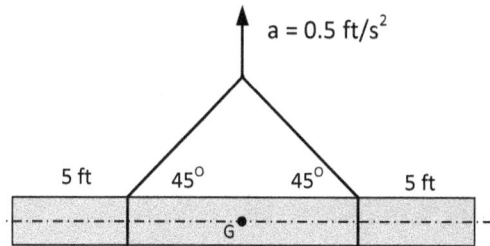

Fig. E6.9 A heavy square bar is being lifted with an acceleration of 0.5 ft/s².

Solution:

Prepare a free body diagram of the entire bar, as shown in Fig. E6.9a.

Fig. E6.9a Free body diagram of the bar as it is lifted.

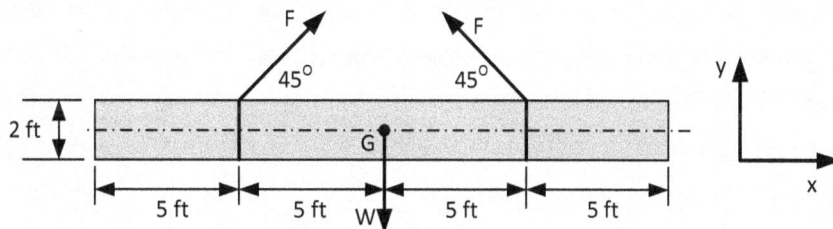

We can determine the magnitude of the forces F by using $\Sigma F = m\, a$.

$$\Sigma F_y = 2F \sin 45^O - W = m\, a = 1.414\, F - 10,000 = (10,000/32.17)(0.5) = 155.4 \text{ lb.}$$

$$F = (1/1.414)(10,000)(1 + 0.5/32.17) = 7,182 \text{ lb} \qquad \text{(a)}$$

To determine the moment M_G, we section the bar and prepare another free body diagram, as shown in Fig. E6.9b.

Fig. E6.9b. A free body diagram of the right half of the square bar, showing the inertia force ma that adds to the weight of the bar.

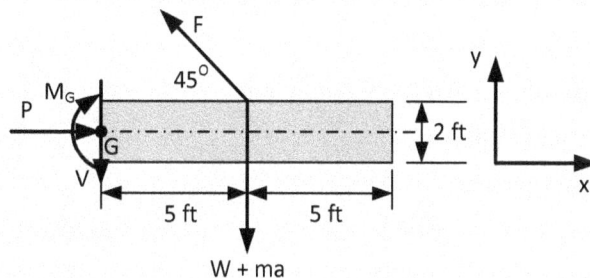

By adding the inertia force vector **ma**, we can consider the half of the bar to be in static equilibrium and write:

$$\Sigma M_G = 0 \qquad \text{(b)}$$

Then $\Sigma M_G = 0$ yields:

$$F \sin 45^O (5) + F \cos 45^O (1) - 5{,}000(1 + 0.5/32.17)(5) - M_G = 0$$

and

$$M_G = 7{,}182(0.7071)(6) - 25{,}000(1.0155) = 30{,}471 - 25{,}388 = 5{,}083 \text{ ft-lb} \qquad (c)$$

Note the shear force V and the normal force P, shown in Fig. E6.9b, was not considered in writing $\Sigma M_G = 0$, because they pass through the center of gravity and their moment arm relative to point $G = 0$.

EXAMPLE 6.10

A bar with a mass of 7 kg is supported by an L bracket, with a mass of 30 kg. The entire assembly is accelerating to the left, with an acceleration a = 2.5 m/s². The surface, upon which the bracket moves, is friction free as is the surface at point B. Determine the force F that moves the entire assembly and the contact force at point B.

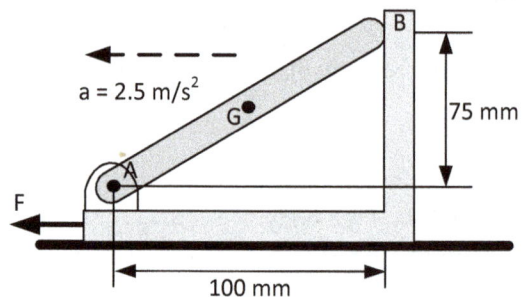

Fig. E6.10 A bracket and bar assembly is accelerating
to the left with a = 2.5 m/s².

Solution:

The entire assembly has a mass m = 30 + 7 = 37 kg. Writing the scalar equation of motion in the horizontal direction gives:

$$\Sigma F: \qquad F = ma = 37\,(2.5) = 92.5 \text{ N} \qquad (a)$$

To determine the contact force B_x, we draw a free body diagram of the bar, as shown in Fig. E6.10a:

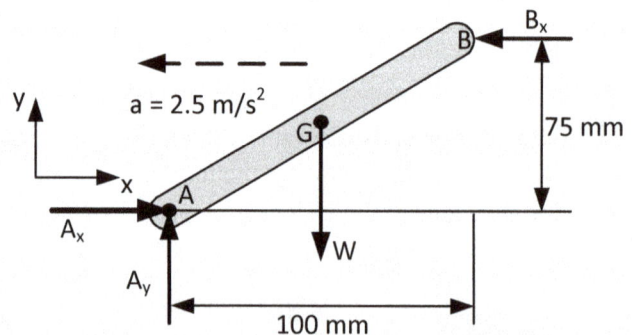

Fig. E6.10a Free body diagram of the bar.

Write the three scalar equations of motion and in the process calculate the forces A_x and A_y at the pin.

$$\Sigma F_y = A_y - W = 0 \qquad\qquad A_y = W = 7(9.807) = 68.65 \text{ N} \qquad (b)$$

$$\Sigma F_x = A_x - B_x = ma = 7(-2.5) = -17.5$$

and

$$A_x = B_x - 17.5 \qquad \text{(c)}$$

$$\Sigma M_G = 37.5(A_x + B_x) - A_y(50) = 0 \qquad \text{(d)}$$

Substitute Eqs. (b) and (c) into Eq. (d) and simplify to obtain:

$$37.5(2B_x - 17.5) = 68.65\ (50) = 3433\ \text{N} \qquad \text{(e)}$$

$$B_x = 54.52\ \text{N}$$

EXAMPLE 6.11

A robotic cart is transporting a large carton that is 2 m high and 0.5 m square. The cart is initially at rest and then is subjected to a constant acceleration a, as shown in Fig. E6.11. The cart has a high friction surface and the carton cannot slide. Determine the minimum time for the cart to achieve a velocity of 3 m/s.

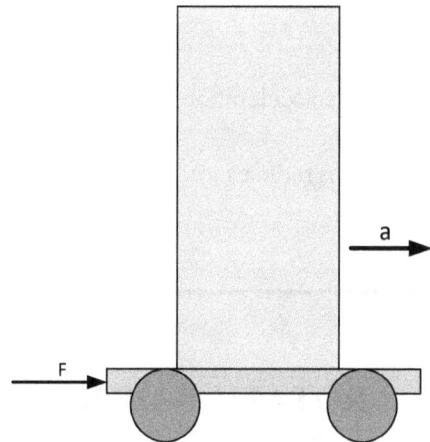

Fig. E6.11 A carton with a weight W is initially at rest.

Solution:

To achieve the velocity of 3 m/s in a minimum amount of time, the acceleration must be maximized. However, if the acceleration is too large the carton will tip over. Let's analyze this situation by constructing a free body diagram of the carton, as shown in Fig. E6.11a.

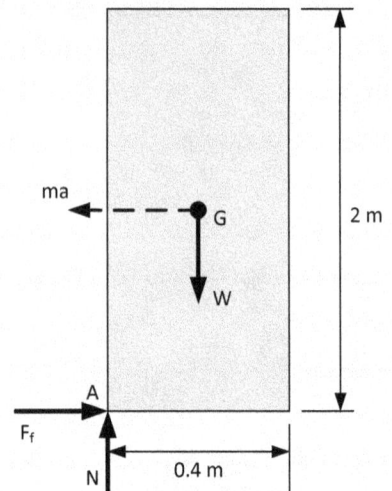

Fig. E6.11a Free body diagram of the carton with the inertia force
vector ma and the forces at the tipping point A.

Take moments about point A to obtain:

$$\Sigma M_A = 0.2\ W - 1.0\ ma = 0 \tag{a}$$

Solving Eq. (a) for the acceleration gives:

$$a = 0.2\ W/m = 0.2Wg/W = 0.2\ g = 0.2(9.807) = 1.961\ \text{m/s}^2 \quad \text{(Maximum)} \tag{b}$$

Recall:

$$v = v_0 + at \tag{c}$$

Because the acceleration is a constant.

Equation (c) reduces to: $\qquad v = at \qquad$ because $v_0 = 0$ $\qquad$ (d)

$$t = v/a = 3/1.961 = 1.530\ \text{s} \tag{e}$$

6.7   ROTATION OF RIGID BODIES

In this section, we will consider rotation of a plane rigid body about a fixed axis. The body resides in the
x-y plane and the axis of rotation is normal to this plane and in the z direction. We divide the topic into
two parts; (1) the rotation about the body's centroid and (2) rotation about an arbitrary, but fixed point
within the body.

6.7.1 Centroidal Rotation

Consider the plane rigid body shown in Fig. 6.3 that is subjected to a system of forces. Because the point
G is fixed, we can write:

$$\Sigma \mathbf{F} = m\mathbf{a_G} = 0 \tag{a}$$

The forces create a moment M_G that is given by:

$$\mathbf{M_G} = \Sigma(\mathbf{r_G} \times \mathbf{F}) \hspace{4cm} \text{(b)}$$

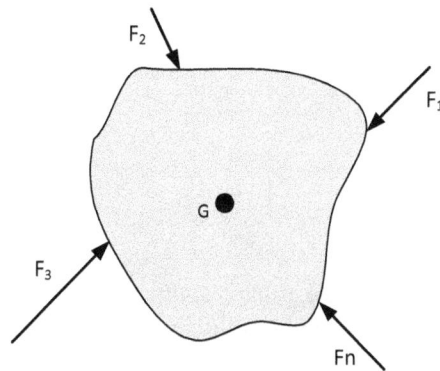

Fig. 6.4 A plane rigid body subjected to a system of forces.

The moment vector $\mathbf{M_G}$ produces an angular acceleration of the body given by:

$$\mathbf{M_G} = I_G\ \boldsymbol{\alpha} \hspace{4cm} \text{(6.15)}$$

The plane body in Fig. 6.4 may be represented by a dynamically equivalent sketch as:

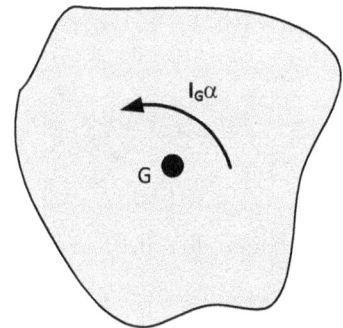

Fig. 6.5 The system of forces ΣF that produce the moment M_G, which in turn produces an angular acceleration α. The forces and moment are not represented in this figure, but the resulting angular acceleration term $I_G\alpha$ is shown.

Recall that the components of the acceleration are the tangential and normal components a_t and a_n, respectively.

$$a_t = r\ \alpha \hspace{4cm} \text{(6.16)}$$

$$a_n = r\ \omega^2 \hspace{4cm} \text{(6.17)}$$

where r locates some point within the body.

Let's consider two examples, to demonstrate the method used in analyzing rotation about a fixed centroidal axis.

EXAMPLE 6.12

A thin disk with a radius r = 0.5 m and a mass of 30 kg is supported at its center, as shown in Fig. E6.12. Attached to its perimeter is a cord that connects with a cylinder, which has a mass of 15 kg. The disk rotates about an axis through its centroid without friction. Determine the angular acceleration α of the disk.

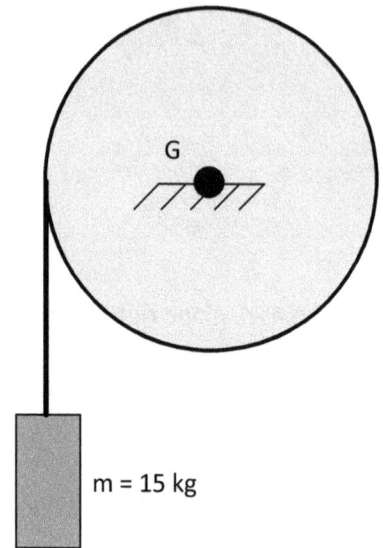

Fig. E6.12 A disk rotates about an axis through G due to the moment produced by the weight of the cylinder.

Solution:

Prepare free body diagrams of the cylinder and the disk.

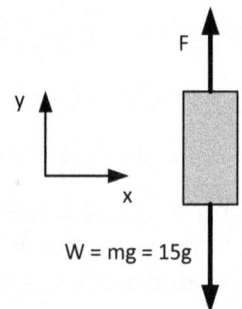

It is evident that the free body diagram of the cylinder yields:

$$\Sigma F_y = F - 15g = -15\,a_y \qquad (a)$$

Note that a_y is the tangential acceleration, where the cable leaves the disk and is acting in the negative y direction:

$$a_y = r_G\,\alpha = 0.5\,\alpha \qquad (b)$$

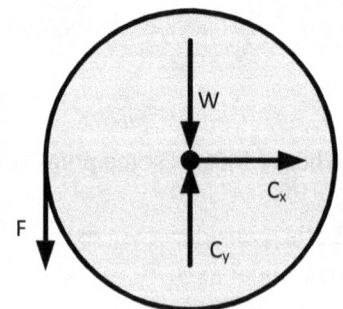

Free body diagram of the disc.

Next examine the free body diagram of the disk and note that ΣF_x and ΣF_y do not yield useful information, because C_x and C_y are both unknown. However ΣM_G does produce an important equation.

$$\Sigma M_G = Fr = I_G\,\alpha \qquad (c)$$

Reference Table 6.1 and write the relation for I_G of the disk as:

$$I_G = \tfrac{1}{2}\, mr^2 = \tfrac{1}{2}\,(30)(0.50)^2 = 3.75 \text{ kg-m}^2 \qquad\qquad\text{(d)}$$

Substitute Eq. (d) into Eq. (c) to obtain:

$$\Sigma M_G = 0.5F = 3.75\,\alpha \qquad\qquad \text{and} \qquad\qquad F = 7.5\,\alpha \qquad\qquad\text{(e)}$$

Substituting Eqs. (b) and (e) into Eq. (a) yields:

$$7.5\,\alpha - 15\,(9.807) = -15\,(0.5)\alpha$$

Solving for α yields:

$$\alpha = 147.1/15 = 9.807 \text{ rad/s}^2 \qquad\qquad\text{(f)}$$

From Eq. (e) we have:

$$F = 7.5\,\alpha = 7.5(9.807) = 73.55 \text{ N} \qquad\qquad\text{(g)}$$

EXAMPLE 6.13

A double disk with radii of 1.0 and 2.0 m, shown in Fig. E6.13, rotates about a fixed axis through point G. Cables attached to the disk support weights with masses of 60 and 100 kg. The radius of gyration of the disk is 1.5 m and its mass is 200 kg. Determine the forces exerted by the two cylindrical masses and the angular acceleration of the double disk.

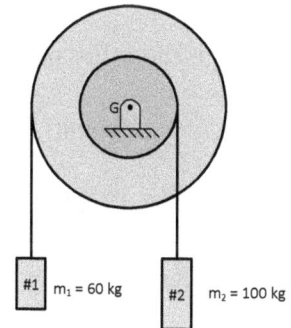

Fig. E6.13 The double disk rotates about the fixed centroidal axis at point G.

Solution:

Draw three free body diagrams and write the motion equations for each of the bodies:

Weight #1 is accelerating downward, because the moment due to m_1 exceeds the moment due to m_2. The double disk is rotating counterclockwise, and the acceleration a_1 of the mass m_1 is negative (in the minus y direction):

$$\Sigma F_y = F_1 - W_1 = ma_1 \qquad\qquad\text{(a)}$$

Note that the angular acceleration of the double disk is in the counter clockwise direction and positive. Hence:

$$a_1 = -r_1\,\alpha = -2.0\alpha$$

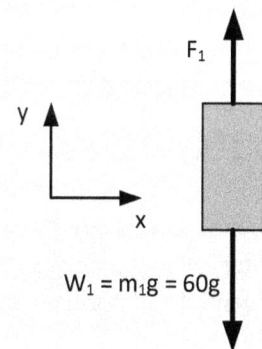

Then

$$F_1 - 60g = 60(-2.0\alpha) = -120\alpha$$

$$F_1 = -120\,\alpha + 588.4 \qquad\qquad (b)$$

Weight #2 is accelerating in the positive y direction; hence, a_2 is positive and we write:

$$\Sigma F_y = F_2 - W_2 = ma_2 \qquad\qquad (c)$$

Note:

$$a_2 = r_2\,\alpha = 1.0\,\alpha$$

Then

$$F_2 - 100g = 100(1.0\,\alpha) = 100\,\alpha$$

$$F_2 = 100\,\alpha + 980.7 \qquad\qquad (d)$$

For the double disk:

$$\Sigma M_G = F_1\,r_1 - F_2\,r_2 = I_G\,\alpha \qquad\qquad (e)$$

We find I_G from:

$$I_G = mk^2 = 200\,(1.5)^2 = 450 \text{ kg-m}^2$$

where k is the radius of gyration of the double disk.

Substituting the results from Eqs. (b) and (d) into Eq. (e) gives:

$$(-120\,\alpha + 588.4)(2.0) - (100\,\alpha + 980.7)(1.0) = 450\,\alpha$$

This reduces to:

$$790\,\alpha = 196.1 \qquad\qquad \text{or} \qquad\qquad \alpha = 0.2482 \text{ rad/s}^2$$

Substituting the result for α into Eqs. (b) and (d) gives:

$$F_1 = -120(0.2482) + 588.4 = 558.6 \text{ N}$$

$$F_2 = 100(0.2482) + 980.7 = 1{,}006 \text{ N}$$

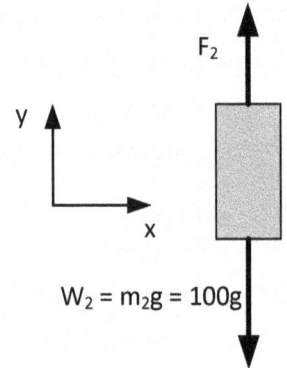

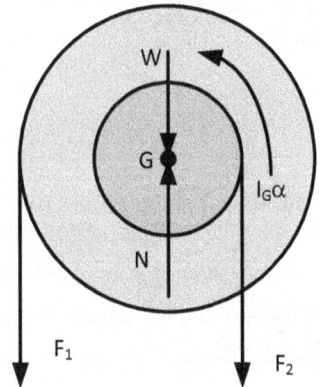

6.7.2 Non Centroidal Rotation

When a plane body rotates about a fixed axis, which does not pass through the centroid of the body, the center of mass moves along a circular path, as shown in Fig. 6.5. With non-centroidal rotation, the tangential and normal components of acceleration are

$$a_G)_t = \alpha\,r_G \qquad\qquad (6.18)$$

$$a_G)_n = \omega^2\, r_G \qquad\qquad (6.19)$$

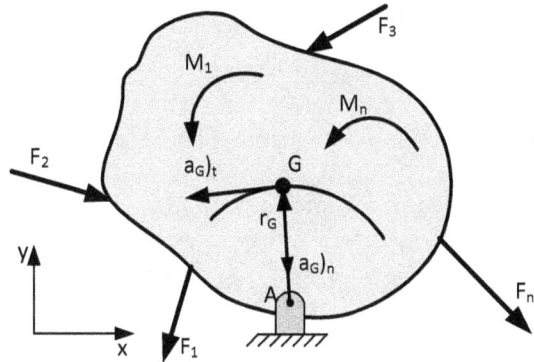

Fig. 6.5 A plane rigid body rotating about a fixed non-centroidal axis.

The scalar equations of motion are:

$$\Sigma F_t = m\, a_G)_t = m\, \alpha\, r_G \qquad\qquad (6.20)$$

$$\Sigma F_n = m\, a_G)_n = m\, \omega^2\, r_G \qquad\qquad (6.21)$$

$$\Sigma M_G = I_G\, \alpha \qquad\qquad (6.22)$$

The equations for ΣF_t and ΣF_n usually do not contribute to a solution, because of the unknown reaction forces at the support for the axis of rotation. For this reason, we rewrite the equation for the moments using point A instead of point G. Then Eq. (6.20) becomes:

$$\Sigma M_A = (I_G + m r_G^2)\alpha = I_A\, \alpha \qquad\qquad (6.23)$$

where:

$$I_A = I_G + m r_G^2 \qquad\qquad (6.24)$$

EXAMPLE 6.14

A long slender rod with a length of 1.6 m and a mass of 12 kg is pinned at one end and free at the other end, as shown in Fig. E6.14. The bar is initially vertical and at rest. The bar becomes unstable and begins to rotate in the clockwise direction, about the pin located at point A. Determine the angular acceleration α and velocity ω when the bar has rotated until the angle $\theta = 45^O$.

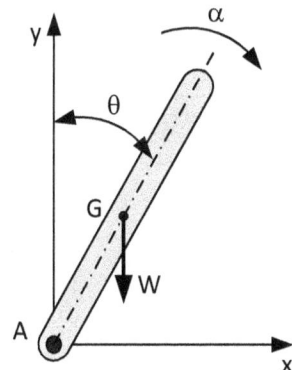

Fig. E6.14 A long slender bar rotating about its end.

Solution:

Equations (6.20) and (6.21) are not useful, because they include the unknown reactive forces at the pin. However, Eq. (6.23) gives:

$$\Sigma M_A = mg\,(L/2)\sin\theta = I_A\,\alpha = (mL^2/3)\,\alpha \qquad (a)$$

From Table 6.1 we note $I_A = mL^2/3$

Then $\qquad \alpha = (3g/2L)\sin\theta = 3(9.807)(0.7071)/[2(1.6)] = 6.501 \text{ rad/s}^2 \qquad (b)$

However, using Eqs. (4.17) and (4.18), we can express $\alpha = d^2\theta/dt^2 = \omega\,d\omega/d\theta$. Substituting this relation into Eq. (b) yields:

$$\omega\,d\omega = (3g/2L)\sin\theta\,d\theta \qquad (c)$$

Integrate Eq. (c) to obtain:

$$\omega^2/2 = -(3g/2L)\cos\theta + C \qquad (d)$$

We determine the integration constant C from the initial conditions that $\omega = \theta = 0$ at $t = 0$.

$$C = 3g/2L \qquad (e)$$

and

$$\omega = [3g/L(1 - \cos\theta)]^{1/2} \qquad (f)$$

Then $\qquad \omega = [3(9.807/1.6)(1 - 0.7071)]^{1/2} = 2.321 \text{ rad/s} \qquad (g)$

EXAMPLE 6.15

The thin disk, shown in Fig. E6.15, rotates in the vertical plane about a pin located at point A, with an angular velocity $\omega = 12$ rad/s. The disk has a radius of 0.4 m and a mass of 40 kg. The angle θ the force F makes with the x axis is 45°. Determine the angular acceleration α and the reaction forces at the supporting pin at A.

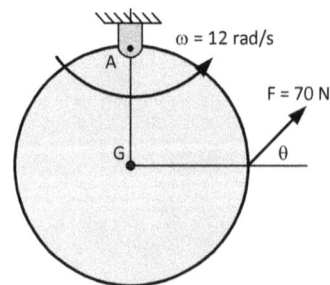

Fig. E6.15 A thin disk supported by a pin connection at point A.

Solution:

To eliminate the reaction forces from the moment equation, we take ΣM about point A and write:

$$\Sigma M_A = F \sin \theta \, r + F \cos \theta \, r = I_A \, \alpha \qquad\qquad \text{(a)}$$

Note that: $I_A = I_G + md^2 = \tfrac{1}{2} m \, r^2 + m \, r^2 = 1.5 \, m \, r^2 = 1.5 \,(40)(0.4)^2 = 9.6 \text{ kg-m}^2 \qquad \text{(b)}$

Substituting numerical values into Eq. (a) gives:

$$70(0.7071)(0.4) + 70(0.7071)(0.4) = 9.6 \, \alpha$$

and

$$\alpha = 39.60/9.6 = 4.125 \text{ rad/s}^2 \qquad\qquad \text{(c)}$$

To determine the reaction forces at the pin located at point A, draw the free body diagram of the disk.

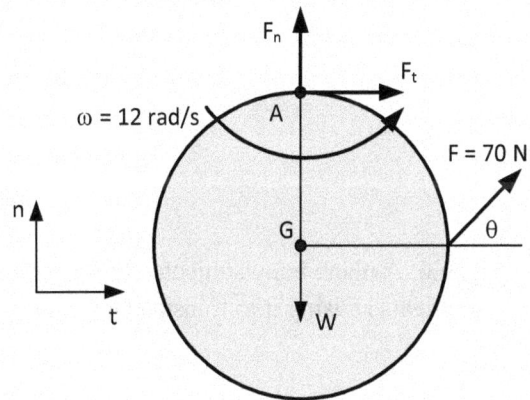

Fig. 6.15a A free body diagram of the disk.

Next write the equations of motion for the disk as:

ΣF_n: $F_n + F \sin \theta - W = ma_G)_n \qquad\qquad \text{(d)}$

Note: $a_G)_n = r \, \omega^2 = (0.4)(12)^2 = 57.6 \text{ m/s}^2$

$\quad F_n = ma_G)_n - F \sin \theta + W = 40(57.6) - 70(0.7071) + 40(9.807) = 2{,}647 \text{ N} \qquad \text{(e)}$

ΣF_t $F_t + F \cos \theta = ma_G)_t \qquad\qquad \text{(f)}$

Note: $a_G)_t = r \, \alpha = (0.4)(4.125) = 1.65 \text{ m/s}^2$

$\quad F_t = ma_G)_t - F \cos \theta = 40(1.65) - 70(0.7071) = 16.50 \text{ N} \qquad \text{(g)}$

6.8   TRANSLATION AND ROTATION OF RIGID BODIES

In the general case of plane motion of rigid bodies, we observe the body rotates and translates. An example is a baseball that is translating when it travels from the pitcher to the batter. However, the ball also rotates due the spin the pitcher initiates, causing the baseball to curve as it approaches the batter. We show a rigid body subjected to both forces and moments in Fig. 6.6. The angular velocity ω and acceleration α are positive in the counterclockwise direction. The center of gravity is identified with the letter G. The equations of motion are referenced to an x-y axes and the moments are taken about the center of gravity.

$$\Sigma F_x = m \, a_G)_x \tag{6.25a}$$

$$\Sigma F_y = m \, a_G)_y \tag{6.25b}$$

$$\Sigma M_G = I_G \, \alpha \tag{6.25c}$$

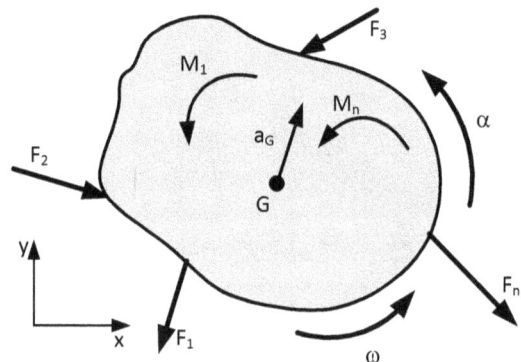

Fig. 6.6 A plane body subjected to forces and moments causing it to translate and rotate.

In some cases, it may be useful to determine the moments about some point removed from the center of mass to avoid unknown forces that complicate the equations of motion. Let's prepare a free body diagram plane body, as shown in Fig 6.7, and consider an arbitrary point Q located in the rigid body.

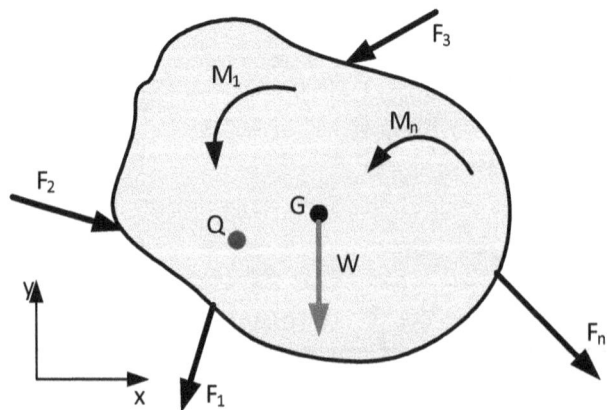

Fig. 6.7 Free body of the plane body showing the forces and moments that produce both translation and rotation.

Next we note that the translation of the point Q and the translation of the center of gravity G are identical with:

$$\Sigma F_x = m \, a_G)_x \tag{6.26a}$$

$$\Sigma F_y = m \ a_G)_y \qquad\qquad (6.26b)$$

However, the angular acceleration about point Q differs from the angular acceleration about the center of gravity G. We express the moment equations about an arbitrary point Q including the inertia forces $m \ a_G)_x$ and $m \ a_G)_y$ due to the linear accelerations.

$$\Sigma M_Q = \Sigma(\mathcal{M}_{\star})_Q = I_Q \ \alpha_Q \qquad\qquad (6.26c)$$

$\Sigma(\mathcal{M}_{\star})_Q$ is given by:

$$\Sigma(\mathcal{M}_{\star})_Q = I_G \ \alpha + \overline{x} \ m \ a_G)_y + \overline{y} \ m \ a_G)_x \qquad\qquad (6.27)$$

where $I_G \ \alpha$ is the inertia moment; $\overline{x} \ m \ a_G)_y + \overline{y} \ m \ a_G)_x$ is the moment about Q due to the inertia forces and $\overline{x}$ and $\overline{y}$ are the distances from point Q to the center of gravity G.

For the final example in Chapter 6, we have selected one that combines kinematics and kinetics. The motion is constrained, and that fact requires methods of kinematics to establish relations for the accelerations. The rigid body is subjected to forces that produce these accelerations and requires the use of the equations of motion associated with kinetics.

EXAMPLE 6.16

A long thin bar with a mass m and a length L is initially at rest on the slope, as shown in Fig. E6.16. Because both of the surfaces are frictionless, the bar will slide down the incline when it is released. Determine the reactions at points A and B and the angular acceleration α of the bar at the instant the bar is released.

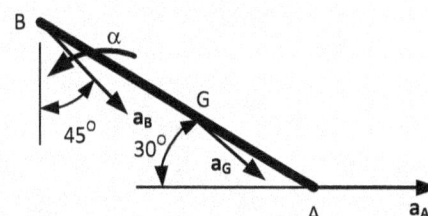

Fig. E6.16 The bar slides down the inclined surface due to its weight and the two friction free surfaces.

Solution:

The motion is constrained by the inclined and level surfaces, along which the bar slides as it moves downward and to the right. We have no initial knowledge of the acceleration of any point on the bar; hence, the equations of motion will not provide a path to the solution for the reaction forces at the contact points A and B. To begin, let's draw a diagram showing the acceleration vectors for points A, B and G on the bar. We know the direction of the acceleration vectors at points A and B, because the bar slides along the inclined and level surfaces.

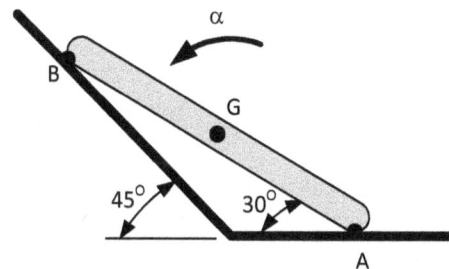

Fig. E6.16a The orientation of the inclined and level surfaces control the direction of the accelerations at point A and B.

From our knowledge of kinematics we can write:

$$\mathbf{a_B} = \mathbf{a_A} + \mathbf{a_{B/A}} \tag{a}$$

As the bar slides down the inclined surface, it is undergoes an angular acceleration α. Then it is evident that:

$$\mathbf{a_{B/A}} = L\,\alpha \tag{b}$$

Because we know the directions of these three vectors, we can draw the vector triangle, shown in Fig. E6.17b.

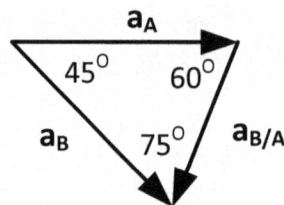

Fig. E6.16b The vector triangle with $\mathbf{a_{B/A}} = L\alpha$

We use the sin law to write:

$$a_B/\sin 60^O = a_{B/A}/\sin 45^O \qquad\qquad a_B = (\sin 60^O/\sin 45^O)\,L\alpha = 1.225\,L\alpha$$
$$\tag{c}$$
$$a_A/\sin 75^O = a_{B/A}/\sin 45^O \qquad\qquad a_A = (\sin 75^O/\sin 45^O)\,L\alpha = 1.366\,L\alpha$$

Following the same procedure we determine a_G.

$$\mathbf{a_G} = \mathbf{a_A} + \mathbf{a_{G/A}} \tag{d}$$

$$\mathbf{a_{G/A}} = (L/2)\,\alpha \tag{e}$$

Draw the vector triangle, shown in Fig. E6.16c.

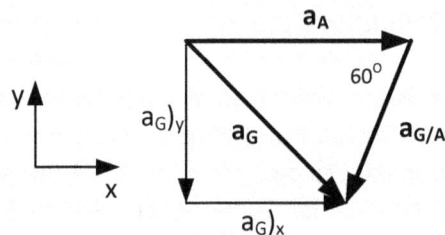

Fig. E6.16c The vector a has been resolved into its x and y components.

From Fig. E6.17c it is evident that:

$$a_G)_y = -\,a_{G/A}\sin 60^O = -\,L/2\,\alpha\sin 60^O = -\,0.4330\,L\alpha$$
$$\tag{f}$$
$$a_G)_x = a_A - a_{G/A}\cos 60^O = 1.366\,L\alpha - L/2\,\alpha\cos 60^O = 1.116\,L\alpha$$

Because the values of the acceleration components in the x and y directions are known, we can write the equations of motion. Begin with a free body diagram of the bar:

Fig. E6.16d Free body diagram of the bar:

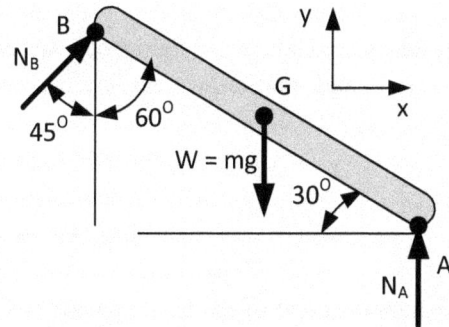

Writing the equations of motion yields:

ΣM_G: $N_A (L/2) \cos 30^O - N_B (L/2) \cos (105^O - 90^O) = I_G \alpha = (1/12)mL^2 \alpha$

$$0.4330 N_A - 0.4830 N_B = 0.0833 \, m \, L\alpha \qquad (g)$$

ΣF_y: $N_A + N_B \cos 45^O - mg = ma_G)_y$

$$N_A + 0.7071 N_B - mg = -0.4330 \, mL\alpha \qquad (h)$$

ΣF_x: $N_B \sin 45^O = ma_G)_x = 1.116 \, mL\alpha$

$$N_B = 1.116 \, L\alpha \, m/0.7071 = 1.578 \, mL\alpha \qquad (i)$$

Substitute Eq. (i) into Eq. (h) to determine N_A as a function of α.

$$N_A + 0.7071(1.578) \, mL\alpha - mg = -0.4330 \, mL\alpha$$

$$N_A = mg - (1.116 + 0.4330)mL\alpha$$

$$N_A = m(g - 1.549 \, L\alpha) \qquad (j)$$

Substitute Eq. (i) and Eq. (j) into Eq. (g) to determine α in terms of g and L.

$$0.4330 \, m(g - 1.549L\alpha) - 0.4830 \, (1.578 \, mL\alpha \,) = 0.0833 \, m \, L\alpha$$

$$0.4330g = (0.0833 + 0.6707 + 0.7622)L\alpha$$

Then:

$$\alpha = 0.4330g/1.516 \, L = 0.2856 \, g/L \qquad (k)$$

$$N_A = m[g - 1.549(0.2856 \, g)] = 0.5576 \, mg \qquad (l)$$

$$N_B = 1.578(0.2856) \, mg = 0.4507 \, mg \qquad (m)$$

6.9   SUMMARY

This chapter deals with rigid body motion including both translation and rotation. To analyze both translation and rotation, we employ two equations of motion:

$$\Sigma \mathbf{F} = m\mathbf{a} \qquad \text{for translation} \qquad (6.1)$$

$$\Sigma \mathbf{M} = I\boldsymbol{\alpha} \qquad \text{for rotation} \qquad (6.2)$$

The mass moment of inertia may be is referenced to specified axes as:

$$I_z = \int_m r^2 dm = \int_m (x^2 + y^2)dm \qquad (6.3a)$$

$$I_x = \int_m r^2 dm = \int_m (y^2 + z^2)dm \qquad (6.3b)$$

$$I_y = \int_m r^2 dm = \int_m (x^2 + z^2)dm \qquad (6.3c)$$

The mass moment of inertia I in terms of a volume element dV is:

$$I = \rho \int_V r^2 dV \qquad (6.4)$$

where ρ is the mass density of the material used in fabricating the rigid body and $dm = \rho dV$.

The parallel axis theorem is:

$$I_z = \overline{I}_z + m(\overline{x}^2 + \overline{y}^2) \qquad (6.5)$$

where $\overline{I}_z = \int_m [x_G^2 + y_G^2]dm$ is the mass moment of inertia about the centroidal z axis.

$$I_x = \overline{I}_x + m(\overline{z}^2 + \overline{y}^2) \qquad (6.6)$$

$$I_y = \overline{I}_y + m(\overline{x}^2 + \overline{z}^2) \qquad (6.7)$$

$$I_z = \overline{I}_z + m\,d^2 \qquad (6.8)$$

where $\qquad\qquad\qquad\quad d = \sqrt{(\overline{x}^2 + \overline{y}^2)} \qquad (6.9)$

In the previous treatment of the translation of rigid bodies, the distribution of the mass of the bodies was not considered. To consider the distribution of the mass of the body we rewrite Eq. (6.1) as:

$$\Sigma \mathbf{F} = m\,\mathbf{a}_G \qquad (6.12)$$

where the acceleration $\mathbf{a}_G$ refers to the center of mass of the rigid body.

The scalar versions of the translation equations of motion as:

$$\Sigma F_x = m\ a_G)_x \qquad\qquad (6.13a)$$

$$\Sigma F_y = m\ a_G)_y \qquad\qquad (6.13b)$$

If the rigid body is translating without rotation:

$$\Sigma M_G = 0 \qquad\qquad (6.13c)$$

With curvilinear motion, but without rotation, the scalar version of Eq. (6.12) are:

$$\Sigma F_n = m\ a_G)_n \qquad\qquad (6.14a)$$

$$\Sigma F_t = m\ a_G)_t \qquad\qquad (6.14b)$$

$$\Sigma M_G = 0 \qquad\qquad (6.14c)$$

Equations (6.14) are employed when analyzing rigid bodies in rectilinear motion.

For centroidal rotation of a plane rigid body subjected to a system of forces, we can write:

$$\Sigma \mathbf{F} = m\mathbf{a} = 0$$

The forces create a moment M_G that is given by:

$$\mathbf{M_G} = \Sigma \mathbf{r_G} \times \mathbf{F}$$

The moment vector $\mathbf{M_G}$ produces an angular acceleration of the body that is given by:

$$\mathbf{M_G} = I_G\ \boldsymbol{\alpha} \qquad\qquad (6.15)$$

Due to rotation, the tangential and normal components of the acceleration are:

$$a_t = r\ \alpha \qquad\qquad (6.16)$$

$$a_n = r\ \omega^2 \qquad\qquad (6.17)$$

With non-centroidal rotation, the tangential and normal components of acceleration are

$$a_G)_t = \alpha\ r_G \qquad\qquad (6.18)$$

$$a_G)_n = \omega^2\ r_G \qquad\qquad (6.19)$$

When a plane body rotates about a non-centroidal axis, the center of mass moves along a circular path and the scalar equations of motion are:

$$\Sigma F_t = m \ a_G)_t = m \ \alpha \ r_G \tag{6.20}$$

$$\Sigma F_n = m \ a_G)_n = m \ \omega^2 \ r_G \tag{6.21}$$

$$\Sigma M_G = I_G \ \alpha \tag{6.22}$$

The equation for the moments about point A instead of point G yields:

$$\Sigma M_A = (I_G + m r_G^2)\alpha = I_A \ \alpha \tag{6.23}$$

where:

$$I_A = I_G + m r_G^2 \tag{6.24}$$

In the general case of plane motion of rigid bodies, we observe the body rotates and translates. The equations of motioned are referenced to an x-y axes and the moments are taken about the center of gravity.

$$\Sigma F_x = m \ a_G)_x \tag{6.25a}$$

$$\Sigma F_y = m \ a_G)_y \tag{6.25b}$$

$$\Sigma M_G = I_G \ \alpha \tag{6.25c}$$

The equations change if we are concerned with the motion of a point Q location some distance from the center of gravity G. The translation of the point Q and the translation of the center of gravity G are identical with:

$$\Sigma F_x = m \ a_G)_x \tag{6.26a}$$

$$\Sigma F_y = m \ a_G)_y \tag{6.26b}$$

However, the angular acceleration about point Q differs from the angular acceleration about the center of gravity G. We express the moment equations about an arbitrary point Q including the inertia forces $m \ a_G)_x$ and $m \ a_G)_y$ due to the linear accelerations as:

$$\Sigma M_Q = \Sigma(\mathcal{M}_k)_Q = I_Q \ \alpha_Q \tag{6.26c}$$

$\Sigma(\mathcal{M}_k)_Q$ is given by:

$$\Sigma(\mathcal{M}_k)_Q = I_G \ \alpha_G + \overline{x} \ m \ a_G)_y + \overline{y} \ m \ a_G)_x \tag{6.27}$$

where ΣM_Q is due to the applied forces and moments; $\overline{x} \ [m \ a_G)_y] + \overline{y} \ [m \ a_G)_x]$ is the moment about Q due to the inertia forces. Note that $\overline{x}$ and $\overline{y}$ are the distances from point Q to the center of gravity G.

CHAPTER 7

WORK AND ENERGY: PARTICLES

INTRODUCTION

In this chapter you will be introduced to three forms of work that is often called energy. First there is work that is performed when a force F moves through some distance s. If the force is constant, determining the work is a simple multiplication of F × s. However, if the force varies over the distance s it is necessary to integrate to determine the work performed. When a force is acting on a particle that moving along a curvilinear path, the work performed is determined by an integration that accounts for the magnitude of the force and the angle it makes with the curvilinear path.

You will learn to determine the work that is required to compress or extend a spring. This work can be recovered when the deformed spring is released.

Work is also required to change the elevation of a weight W. You should recognize that the work performed, due to a change in elevation of a particle in a gravitational g field, is equal to the potential energy E_P of the particle. Potential energy can be gained or lost by changing the height of the particle.

You will learn that the kinetic energy of a particle is ½ mv^2. You will also learn that when energy is conserved you are able to write equations equating energy before and after some event. .

We introduce a new quantity called power P that is defined as the rate of change of work E_W with respect to time. Recognize that power is a scalar quantity and you are concerned only with its magnitude. In the SI system of units, power P is expressed in watts **W** and in the U. S. Customary system of units the power P is expressed in horsepower HP. You will learn that efficiency is defined the ratio of the power output divided by the power input.

CONCEPT PROBLEM

A common method for balancing load for electrical power plants is to employ pumped storage. This method involves pumping large quantities of water from a river or pond at a low elevation to a storage pond located on a nearby hill or mountain. The pumping is performed during the late night and early morning hours when the demand is low. During the high demand hours late in the day and early evening, the water from the storage pond is transferred via a penstock down to the power plant, where it is used to turn turbines and generators producing needed electricity. This balancing load method employs work (pumps) to increase the potential energy of a large quantity of water, when the demand from customers is low. Then the process is reversed and the potential energy of the water is reduced to produce work (turning the turbines and generators) to produce electricity when the demand is high. In this process, energy is lost and load balancing requires capital investment and operating costs. However, it is less costly than dumping power that must be generated and distributed regardless of the system demand.

To illustrate the concept of load balancing consider the power plant located beside a river on the adjacent bottom land, as shown in the figure below:

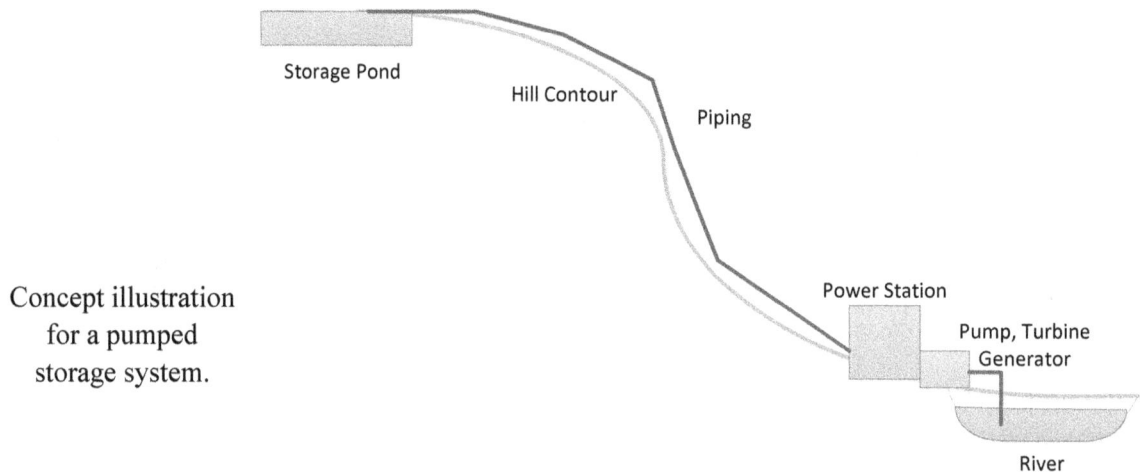

Concept illustration for a pumped storage system.

Storage Pond

Hill Contour

Piping

Power Station

Pump, Turbine Generator

River

A 1,200 ft high hill is adjacent to the power plant. A large storage pond has been constructed on top of the hill, as indicated in the figure above. From 10 PM to 6 AM water is pumped at the rate of 1,000 GPM (gallons per minute) from the river to the storage pond. From 3 PM to 7 PM water is transferred at the rate of 2,000 GPM (gallons per minute) from the storage pond to the power station. This flow is directed into the turbine that in turn drives a generator producing power to meet the higher afternoon demand. The water output from the turbine is returned to the river.

Let's begin our analysis of the pumped storage system by determining the work required to pump one gallon of water 1,200 ft from the river to the storage pond. Note that one gallon of water weighs 8.336 lb and that work is the product of force (weight) through a distance. Then the work to lift one gallon of water is:

$$E_{W1} = F\ d = 8.336\ (1,200) = 10,000 \text{ ft-lb} \qquad \text{(a)}$$

Now that we have determined the work required to lift one gallon of water into the storage pond, let's determine the number of gallons that the personnel at the power station will store during the eight hour pumping period.

$$N = \text{GPM (time)} = 1,000\ (8)(60) = 48 \times 10^4 \text{ gallon} \qquad \text{(b)}$$

The work required to pump 48×10^4 gallon 1,200 ft from the river to the storage pond is determined from Eqs. (a) and (b) as:

$$E_{W2} = 48 \times 10^4\ (10 \times 10^3\ ) = 480 \times 10^7 \text{ ft-lb} \qquad \text{(c)}$$

The result for the work required in Eq. (c) is an ideal result, because it does not take into account losses due to pump efficiency and friction and turbulence losses in the piping. We will assume that an axial piston pump with an efficiency of 88% is employed to transfer the water. We will also neglect the friction and turbulent losses in the piping. Taking account of the pump efficiency $\mathcal{E}_p$ increases the work required to pump the water to:

$$E_{W3} = E_{W2}/\mathcal{E}_p = 480 \times 10^7/0.88 = 545.5 \times 10^7 \text{ ft-lb} \qquad \text{(d)}$$

The water in the storage pond has potential energy $E_P = E_{W2}$ that can be recovered with suitable machinery. Let's recover this potential energy of the water, during the high demand period. Ideally the work recovered E_{W4} would be equal to E_{W2}. However, the efficiency $\mathcal{E}_{\eta g}$ of the turbine-generator set is only 82%; hence the energy recovered E_{W4} is:

$$E_{W4} = \mathcal{E}_{\eta g} (E_{W2}) = 0.82 (480 \times 10^7) = 393.6 \times 10^7 \text{ ft-lb} \qquad \text{(e)}$$

The results in Eqs. (d) and (e) show that energy recovered is less that the energy expended and the pumped storage system efficiency is:

$$\mathcal{E}_{System} = E_{W4}/E_{W3} = 393.6/545.5 = 0.7213 \qquad \text{(f)}$$

Power companies with suitable terrain and a source of large volumes of water will often expend significant capital to construct pumped storage systems, because of the different rates for electricity as a function of demand. They use cheap electricity to pump and sell more expensive electricity later when the demand is high. Another factor is that reactors, boilers and steam turbines cannot be shut down when demand drops. Indeed, they are only shut down for major maintenance once or twice a year.

DISCUSSION

The demand for electrical power varies throughout the 24 hour day, as shown in the graph to the right. In New England in October, power demand is at a minimum between midnight and about 5 AM. Then the demand ramps up by about 40% by 9 AM and remains nearly constant until 9 PM, when it begins to drop off.

You might ask why the power company does not sell electricity to a power company a couple of time zones away where demand is still strong. We have transmission lines for transporting power at very high voltages. Even with high voltages resistance losses in the transmission lines limit the distance power can be economically transmitted. To deal with the problem of variable demand power companies offer variable rates to large industrial users. They might charge $0.08 for a kW of

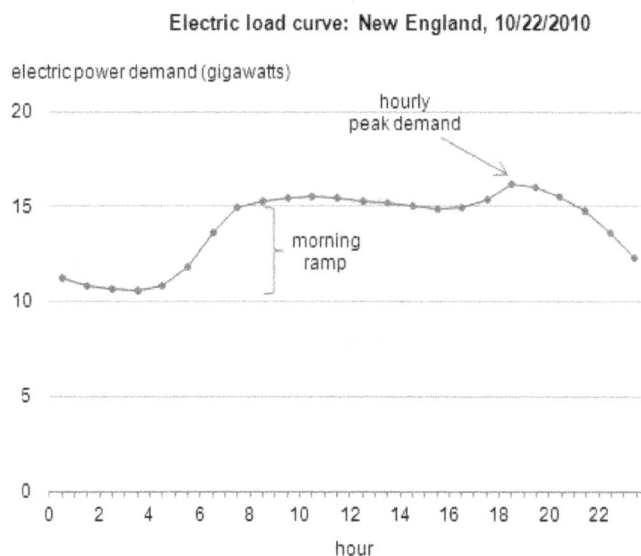

Electric load curve: New England, 10/22/2010

electric power demand (gigawatts)

electricity delivered after 10 PM and until 4 AM and then increase the cost to $0.15 kW for the remaining time period.

Many power companies have begun to install solar farms, where the photons from the sun are converted into electricity. Fortunately the sun, when it does shine, shines during the period when demand is high. Power generated by the winds is less predictable and when the windmills generate power in the darkness they may add power to the grid when it is not needed.

Gas turbines similar to those used to power airplanes are different from steam turbines. They can be started with little or no warm up time and they can be shut down without issues. For power generation, the turbines are fueled with natural gas. The gas is ignited and the combustion products drive a gas turbine which in turn drives a generator to produce electricity. Often the gasses exiting the turbine are hot enough to heat water in a boiler and a second stage steam turbine drives a second generator to produce more power. The use of the exhaust heat from the first (gas) turbine enables relatively high efficiencies for this system.

MORE DISCUSSION

In writing the problem statement, we indicated that an axial piston pump was employed for pumping the water 1,200 ft to the storage pond. You might have wondered why we did not indicate that the water was being pumped with a more common and much less expensive centrifugal pump. The reason is the pressure required to drive the water to a height of 1,200 ft. Let's calculate that pressure.

$$p = \gamma\, h = 62.4\,(1{,}200) = 74{,}880 \text{ lb/ft}^2$$

where $\gamma = 62.4$ /ft^3 is the density of fresh water and h is the height of the water column.

Converting this pressure to lb/in^2 gives:

$$p = 74{,}880/144 = 520 \text{ lb/in}^2$$

A electric motor driven centrifugal pump is depicted in the figure below. Note that the impeller consists of vanes that take the water entering at the center of the pump and with rotation throw the water into the vertical output pipe. This design is effective when the output head (pressure) is relatively low. However, when the head is 100 ft or higher, the efficiency of the centrifugal pump decreases rapidly with increasing head. For this reason, we have selected a positive displacement pump. Of the several positive displacement pumps available, we selected the axial piston pump that is showed in the figure below.

Axial Piston Pump

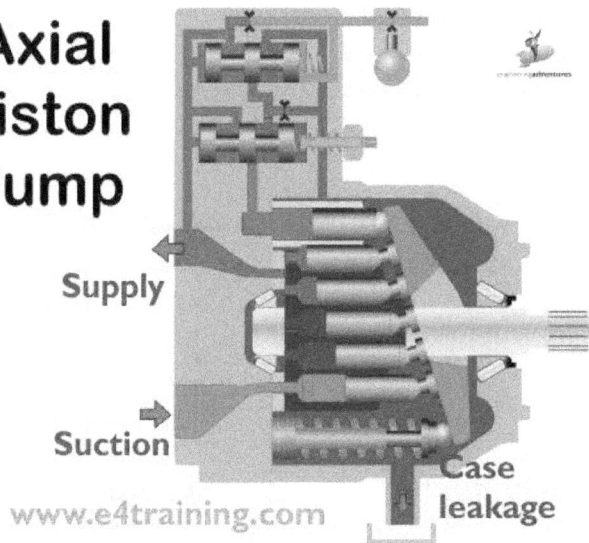

Supply

Suction

Case leakage

www.e4training.com

Axial Piston Pump

Centrifugal Pump

7.1 WORK DUE TO A FORCE ACTING ON A PARTICLE IN LINEAR MOTION

Consider a particle residing on the x axis, as shown in Fig. 7.1. Suppose it is subjected to a constant force F for an increment of time, as it moves from x_0 to x_1. The work E_W performed by the force F is given by:

$$E_W = F \int_{x_1}^{x_2} dx = F(x_1 - x_2) \qquad (7.1)^1$$

Fig. 7.1 A particle moving under the action of a constant force F.

O

F x_1 F x_2 x

Under the action of a constant force, the work performed on the particle increases linearly with the distance the particle moves, as shown in Fig. 7.2.

E_W

Fig. 7.2 The work E_W increases linearly with x due to a constant force F.

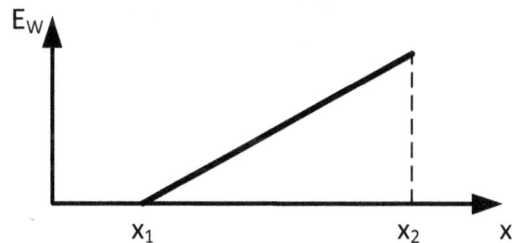

x_1 x_2 x

---

[1] We assigned the symbol E_W for the work performed on a particle to remind the reader that work is a form of energy.

If F is not a constant, but a function of x with F = f(x), then the work performed by f(x) is:

$$E_W = \int_{x_1}^{x_2} f(x)dx \qquad (7.2)$$

To demonstrate the use of Eq. (7.2), consider a spring with a spring rate k, positioned on the x axis as shown in Fig. 7.3. The force required to compress a spring is given by:

$$F = f(x) = kx \qquad (7.3)$$

Substitute Eq. (7.3) into Eq. (7.2) yields:

$$E_W = k\int_{x_1}^{x_2} xdx = (1/2)k\left[x_2^2 - x_1^2\right] \qquad (7.4)$$

Fig. 7.3 Work is required to compress or extend a spring from its free length. This spring is initially compressed from x = 0 to x_1 and then to x_2. Work is recovered from the spring, when it is permitted to expand or compress to its free length.

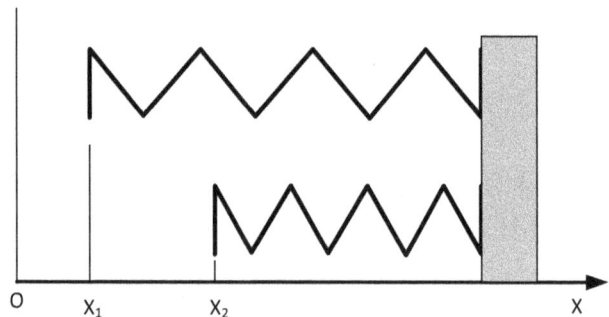

The work required to compress a spring with a spring rate of k is depicted in the graph shown in Fig. 7.4. The work is shown as the gray area that consist of a rectangle and a triangle. The area of the rectangle is $kx_1 (x_2 - x_1)$ and the area of the triangle is $\frac{1}{2}(kx_2 - kx_1)(x_2 - x_1)$. Add these two quantities to confirm Eq. 7.4.

Fig. 7.4 A graphical representation of the work required to compress a spring with a spring rate of k.

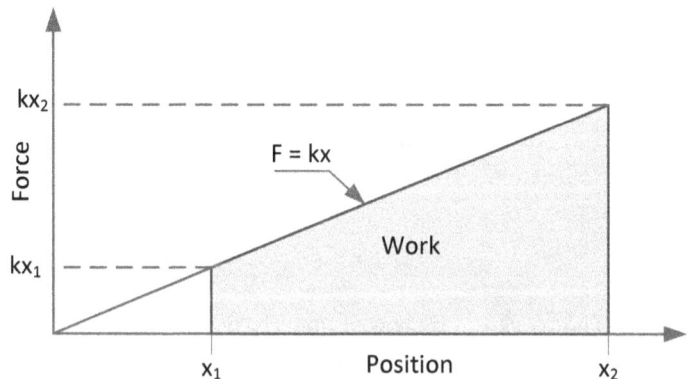

Another example of work, pertains to the change in elevation of particles with a weight W = mg. In Fig. 7.5, we show a particle moving along a curvilinear path from an elevation y_2 to a lower elevation y_1. In the process the particle has lost elevation and hence, it has also lost potential energy. The amount of potential energy lost is equal to the work performed to move the particle between y_2 and y_1.

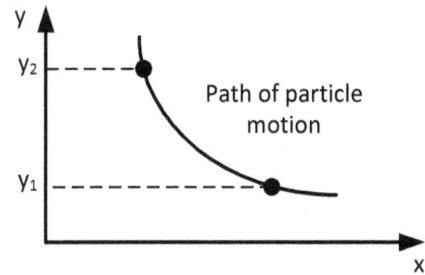

Fig. 7.5 A particle changes elevation as it moves along a curvilinear path and in the process the particle loses potential energy.

Because the particle loses potential energy as it moves to a lower elevation with $y_2 > y_1$, the work performed is negative as indicted by:

$$E_W = mg(y_1 - y_2) = W\,(y_1 - y_2) \tag{7.5}$$

This work is considered as a negative quantity, because the particle loses potential energy, as it moved from a higher to a lower elevation. The work due to the change in elevation of the particle in a gravitational force field is called the potential energy E_P of the particle. Hence we may write:

$$E_W = W\,(y_1 - y_2) = E_P)_1 - E_P)_2 \tag{7.6}$$

The work E_W due to the change in elevation of the weight W is independent of the path of the particle and depends only on the elevations y_1 and y_2.

7.2 WORK DUE TO A FORCE ACTING ON A PARTICLE IN CURVILINEAR MOTION

A more general case of plane motion of a particle involves its movement along a curvilinear path when subjected to an applied force that is not tangential to the path of the particle. We illustrate this case in Fig. 7.6, where the particle moves an incremental distance dr = ds along a curved path. The force F makes an angle θ with the tangent to the curved path during this incremental movement.

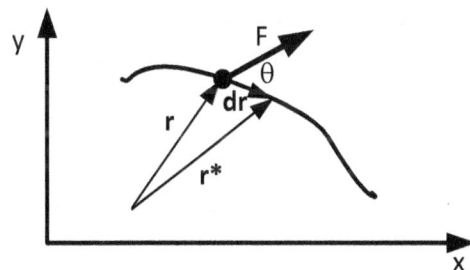

Fig. 7.6 A particle moving along a curved path with an applied force that is not tangent to the vector **dr**.

The incremental work dE_W due to the force is:

$$dE_W = \mathbf{F} \bullet \mathbf{dr} \tag{7.7}$$

where $\bullet$ is the dot product operator. Integrating Eq. (7.7) and treating F as a variable with r, we obtain:

$$E_W = \int_{r_1}^{r_2} \mathbf{F} \bullet \mathbf{dr} = \int_{s_1}^{s_2} F\cos\theta\,ds \tag{7.8}$$

EXAMPLE 7.1

A carton weighing 220 lb, resting on a level floor, is acted upon by a force of 80 lb that makes an angle of 30 degree with the x axis. The friction coefficient $\mu = 0.25$ between the floor and the carton. Find the work accomplished in moving the box a distance of 12 ft.

Fig. E7.1 A box moved along a horizontal floor due to the force F.

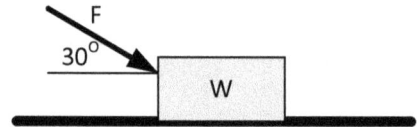

Solution:

Prepare a free body diagram of the carton.

ΣF_y:

$$N - F \sin 30^\circ - W = 0$$

$$N = 80(0.50) + 220 = 260 \text{ lb} \qquad \text{(a)}$$

The friction force is:

$$F_f = \mu N = 0.25(260) = 65 \text{ lb} \qquad \text{(b)}$$

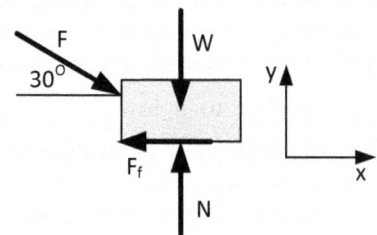

The work E_W performed in moving the cartoon a distance of 12 ft is:

$$E_W = (F \cos 30^\circ - F_f)(x_2 - x_1) = (69.28 - 65)(12 - 0) = 51.38 \text{ ft-lb} \qquad \text{(c)}$$

The portion of the work to overcome the friction force is lost to our local system. The work performed in overcoming friction is converted to heat. From an engineering viewpoint this energy is lost, because we cannot capture the heat energy and convert it to a more useful form of energy. However, in some cases such as the heat lost from a condenser in a power plant can be partially recovered by using it in preheaters to heat air and water entering the boilers.

While the heat is lost to the local system it is not completely lost, because it is added to the global accumulation of heat energy. We define global as the Earth's atmosphere and outer space. Heat from the Earth's surface and its atmosphere is radiated into space; hence, we consider global to refer to both the Earth and space.

Another approach for determining the work performed is to determine the resultant of the force components in the x direction and multiplying that quantity by the distance the object was moved.

ΣF_x:

$$R_x = F \cos 30^\circ - F_f = 80(0.8660) - 65 = 4.28 \text{ lb} \qquad \text{(d)}$$

$$E_W = R_x (x_2 - x_1) = 4.28 (12) = 51.36 \text{ ft-lb} \qquad \text{(e)}$$

EXAMPLE 7.2

A block with a mass of 10 kg is at rest on an inclined plane that makes an angle of 30° with a horizontal plane. A force F = 140 N, which makes an angle of 10° with the x* axis, is applied to the block. The coefficient of friction is $\mu = 0.2$. Determine the work required to move the block a distance of 6 m.

Fig. E 7.2 A block resting on an inclined plane. Note the orientation of the x*-y* axes.

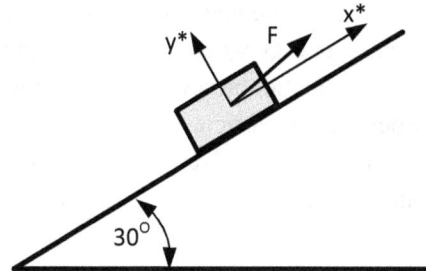

Solution:

Prepare a free body diagram of the block.

ΣF_{y*}: $N - W \cos 30^\circ + F \sin 10^\circ = 0$

$$N = 10(9.807)(0.8660) - 140(0.1736) = 0$$

$$N = 60.62 \text{ N} \qquad \text{(a)}$$

The friction force F_f is:

$$F_f = \mu N = 0.2(60.62) = 12.12 \text{ N} \qquad \text{(b)}$$

The work performed by each force is:

$$E_W = [F \cos 10^\circ - F_f - W \sin 30^\circ](x_2 - x_1) \qquad \text{(c)}$$

$$E_W = [140(0.9848) - 12.12 - 98.07(0.5)](6 - 0) = 76.71(6)$$

$$E_W = 460.3 \text{ N-m} \qquad \text{(d)}$$

The negative quantities in Eq. (c) are due to the forces that oppose the motion of the particle.

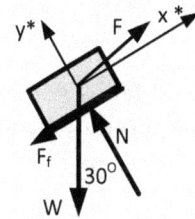

7.3   KINETIC ENERGY

There are several forms of energy. Work is a form of energy that can be added to or subtracted from a local system. Potential energy involves the change in the elevation of a mass or the energy stored in a spring. A system gains potential energy when the mass is raised to a higher level and loses potential energy when the mass is lowered. A system gains potential energy when a spring is compressed and loses potential energy when it returns to its free length.

Kinetic energy, another form of energy and the topic in this section, involves the mass and the velocity of a particle. Let's consider a particle moving with a velocity v along a curvilinear path, as shown in Fig. 7.7. Next let's and write the tangential component for the equation of motion of the particle using Eq. (2.9) to give the relation for the tangential acceleration as.

$$F_t = ma_t = m \, v \, dv/dx \qquad (7.9)$$

Fig. 7.7 A particle with a mass m is moving along a curvilinear path with a velocity v.

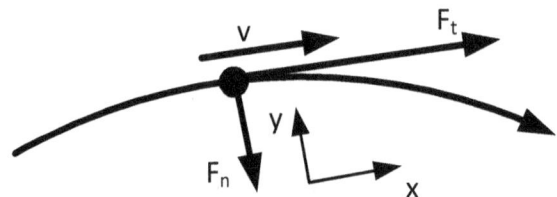

Integrating Eq. (7.9) with respect to x yields:

$$\int_{x_1}^{x_2} F_t dx = m \int_{v_1}^{v_2} v dv \qquad (7.10)$$

Substitute Eq. (7.2) into Eq. (7.10) to obtain:

$$E_W = \tfrac{1}{2} \, m(v_2^2 - v_1^2) \qquad (7.11)$$

We define this quantity as the change in kinetic energy and write:

$$E_k = \tfrac{1}{2} \, m(v_2^2 - v_1^2) \qquad (7.12)$$

It is clear from Eq. (7.12) that the kinetic energy of a particle is always positive, because the velocity term is squared. The term $\tfrac{1}{2} \, mv_2^2$ is the kinetic energy when the particle moves to point 2, and the term $\tfrac{1}{2} \, mv_1^2$ is the kinetic energy when the particle is at point 1. Both of these terms are positive, although the difference may be negative.

We consider that energy is conserved, which enables us to write:

$$E_k)_1 + \Sigma E_W = E_k)_2 \qquad (7.13)$$

The left side of this relation is the initial kinetic energy plus the work that is added or removed from the system. The right side of the relation is the final kinetic energy.

EXAMPLE 7.3

A block, initially at rest with a mass of 12 kg, is subjected to a force F = 750 N, as shown in Fig. E7.3. A compressive spring with a spring rate k = 400 N/m resists the motion. The spring is not compressed initially. The surface upon which the block slides is friction free. Determine the velocity of the block, after it has moved a distance d = 0.8 m.

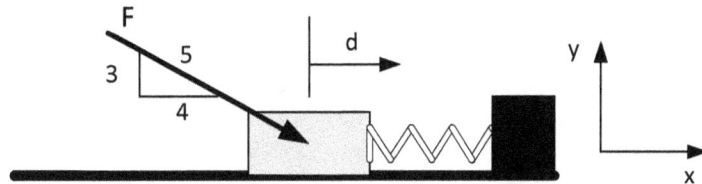

Fig. E 7.3 A local system consisting of
a block and a spring.

Solution:

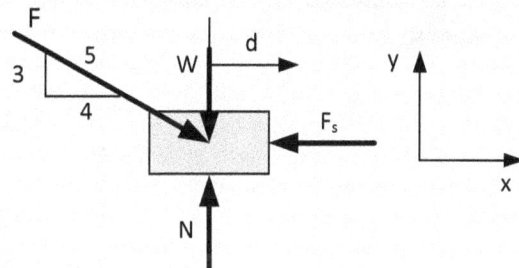

Draw a free body diagram of the block.

The work performed by the force F is:

$E_W)_F = (4/5)F\,d = 0.8(750)(0.8) = 480$ N-m   (a)

The work required to compress the spring is:

$$E_W)_S = \tfrac{1}{2}\,kd^2 = \tfrac{1}{2}\,(400)(0.8)^2 = 128 \text{ N-m} \qquad\qquad (b)$$

The sum of the work is:

$$E_W)_F - E_W)_S = 480 - 128 = 352 \text{ N-m} \qquad\qquad (c)$$

Recall Eq. (7.13):

$$E_k)_1 + \Sigma E_W = E_k)_2$$

$$0 + 352 = \tfrac{1}{2}\,mv^2 \qquad\qquad (d)$$

$$v^2 = 2(352)/12 = 58.67 \qquad\text{and}\qquad v = 7.659 \text{ m/s} \qquad\qquad (e)$$

EXAMPLE 7.4

A race car with a mass of 1,400 kg is traveling at a speed of 120 m/s, when it shuts down its engine and simultaneously deploys its drag chute. If the drag force provided by the chute is given by the force distance diagram that is shown in Fig. E7.4, determine the velocity of the race car after the chute has been deployed for 300 m.

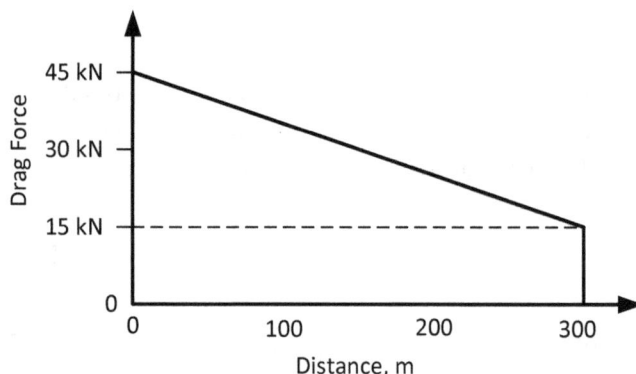

Fig. E7.4 Drag force as a function of distance
with the chute deployed.

Solution:

We will compute the initial kinetic energy of the race car and then subtract the work performed by the drag chute to give the final kinetic energy of the car after it has traveled 300 m.

$$E_k)_1 = \tfrac{1}{2}\,mv_1^2 = \tfrac{1}{2}\,(1{,}400)(120)^2 = 10.08 \times 10^6 \text{ kg-m}^2 \qquad (a)$$

The work performed by the drag chute is the area under the force distance curve, shown in Fig. E7.3.

$$E_W)_D = 15 \times 10^3 (300) + \tfrac{1}{2}\,(30 \times 10^3)(300) = 9.0 \times 10^6 \text{ N-m} \qquad (b)$$

The kinetic energy after travelling 300 m is:

$$E_k)_2 = E_k)_1 - E_W)_D = \tfrac{1}{2}\,mv_2^2 \qquad (c)$$

$$\tfrac{1}{2}\,mv_2^2 = (10.08 - 9.0) \times 10^6 = 1.08 \times 10^6 \qquad (d)$$

$$v^2 = 2(1.08 \times 10^6)/1{,}400 = 1{,}543 \text{ m}^2/\text{s}^2 \qquad v_2 = 39.28 \text{ m/s} \qquad (e)$$

EXAMPLE 7.5

A cart, initially at rest with a mass m = 15 kg, is moved from position #1 to position #2, with a cable that extends from the cart to a small pulley located directly below position #2, as shown in Fig. E7.5. A constant force F = 240 N is applied to the cable to move the cart a distance of 9 m. Determine the velocity of the cart, when it reaches position #2. The surface over which the cart travels is smooth.

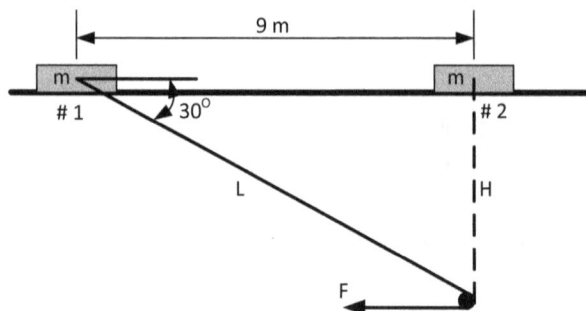

Fig. E7.5 A cart is pulled from position #1 to
position #2 with a cable pulley arrangement.

Solution:

Because the cart is initially at rest $E_k)_1 = 0$.

The work performed by the force F is determined by the length of the cable pulled over the pulley. The total cable length is L and the pulley is located a distance H below position #2. After the mass m is pulled to position #2, the length of the cable is L; hence, (L – H) is the distance the cable was pulled to move the mass m.:

$$E_W = F\,(L - H) \tag{a}$$

where $\qquad L = 9/\cos 30^O = 10.392 \text{ m} \qquad$ and $\qquad H = L \sin 30^O = 10.39(1/2) = 5.196 \text{ m}$

$$E_W = F\,(L - H) = 240(10.392 - 5.196) = 1{,}247 \text{ N-m} \tag{b}$$

$$E_k)_2 = \tfrac{1}{2}\,mv^2 = E_W \tag{c}$$

$$v^2 = 2(1{,}247)/15 = 166.3 \text{ m}^2/\text{s}^2$$

$$v = 12.89 \text{ m/s} \tag{d}$$

7.4   POWER AND EFFICIENCY

7.4.1 Power

We define power P as the rate of change of work E_W with respect to time. When performing work, we seek an engine or motor with a high power rating, because it will perform work in less time than an engine with a lower power rating. We determine the power P by differentiating the work E_W with respect to time:

$$P = dE_W/dt \tag{7.14}$$

Recall Eq. (7.7): $\qquad\qquad\qquad dE_W = \mathbf{F} \bullet d\mathbf{r} \tag{a}$

where $\bullet$ is the dot product operator and $\mathbf{F}$ and $\mathbf{r}$ are vectors.

From Eq. (7.14) and (a), we write:

$$P = \mathbf{F} \bullet d\mathbf{r}/dt \tag{7.15}$$

$$P = \mathbf{F} \bullet \mathbf{v} \tag{7.16}$$

The power is a scalar quantity, because the dot product of two vectors yields a scalar result.

In the SI system, the unit for power P is expressed in watts $\mathbf{W}$ where:

$$1\,\mathbf{W} = 1 \text{ joule/s} = 1 \text{ J/s} = 1 \text{ N-m/s}$$

In the U. S. Customary system, the unit for power P is expressed in horsepower **HP**, where:

$$1\ \textbf{HP} = 550\ \text{ft-lb/s} = 33,000\ \text{ft-lb/min.}$$

7.4.2 Efficiency

Machinery, engines and motors are not perfect. They exhibit friction and electrical loses. Hence, in operation, they require more input power than the power they deliver (output). We define the power output divided by the power input as the efficiency of the device.

$$\mathcal{E} = P_{Out}/P_{In} \qquad\qquad (7.17)$$

The relation for efficiency may also be expressed in terms of energy E_{In} and energy E_{Out} of a system, if both are measured over the same time interval. Hence:

$$\mathcal{E} = E_{Out}/E_{In} \qquad\qquad (7.18)$$

Let's consider a few examples to demonstrate the use of these equations.

EXAMPLE 7.6

A box with a mass m = 28 kg is moving up an inclined plane at a constant velocity of 7.5 m/s in the x* direction, as shown in Fig. E7.6. The friction coefficient $\mu = 0.25$, between the box and the inclined surface. Determine the power developed by the force F.

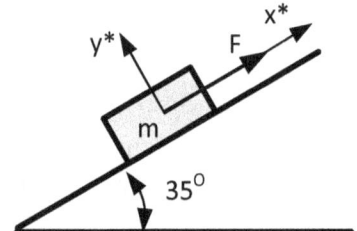

Fig. E7.6 A force F is moving the box up the inclined plane with a constant velocity of 7.5 m/s.

Solution:

Prepare a free body diagram of the block and write the equations of motion in the x* and y* directions.

F_{y*}: $\qquad\qquad N - W \cos 35^{O} = ma_{y*} = 0$

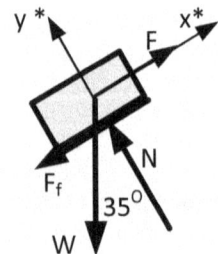

$\qquad\qquad\qquad N = (28)(9.807)(0.8192) = 224.9\ \text{N} \qquad (a)$

$\qquad\qquad\qquad F_f = \mu N = 0.25\,(224.9) = 56.23\ \text{N} \qquad (b)$

ΣF_{x*}: $\qquad\qquad F - F_f - W \sin 35^{O} = ma_x = 0 \qquad (c)$

Note that $a_{x*} = 0$, because the velocity was constant.

$\qquad\qquad F = 56.23 + 28(9.807)(0.5736) = 213.7\ \text{N} \qquad (d)$

The power is given by: $\qquad P = F\,v = 213.7(7.5) = 1,603\ \text{W} = 1.603\ \text{kW} \qquad (e)$

EXAMPLE 7.7

The motor-pulley arrangement, shown in Fig. E7.7, is employed to lift a mass m = 60 kg. The velocity and acceleration at a point on the cable, at this instant in time, is v = 6 m/s and a = 4 m/s². The efficiency of the electric motor is 90%. Determine the power input to the motor.

Fig. E7.7 The motor pulley arrangement for lifting the mass m.

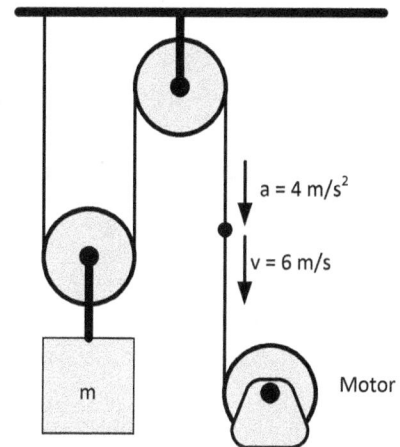

Solution:

Prepare a free body diagram of the mass and pulley:

ΣF_y:

$$2T - W = ma_y \qquad \text{(a)}$$

$$a_y = a/2 = 4/2 = 2 \text{ m/s}^2 \qquad \text{(b)}$$

$$T = \tfrac{1}{2}(60)(2 + 9.807) = 354.2 \text{ N} \qquad \text{(c)}$$

The power output is:

$$P_{Out} = T\,v = 354.2(6) = 2{,}125 \text{ W}$$

The power input is:

$$P_{In} = P_{Out}/\mathcal{E} = 2{,}125/0.9 = 2{,}361 \text{ W} = 2.361 \text{ kW}$$

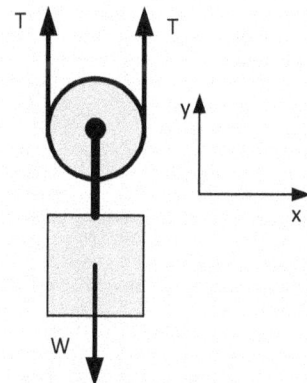

EXAMPLE 7.8

A SUV, with a mass m = 1,800 kg, is initially at rest before it proceeds up a road with a 5° incline, as shown in Fig. E7.8. The velocity increases at a constant rate from 0 to 20 m/s during the 40 second time interval of acceleration. The engine efficiency is 40%. Determine the maximum power output and power input. Also find the average power output and input of the engine. Assume the friction coefficient μ is sufficient that the tires on the SUV do not slip as the vehicle moves up the hill.

Fig. E7.8 A SUV accelerating up a 5° slope.

Solution:

Prepare a free body diagram of the SUV.

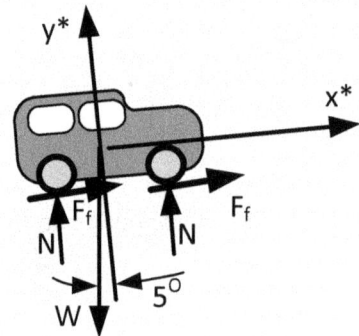

Write the equation of motion for the components in the x* direction:

$$\Sigma F_{x*}: \qquad 2F_f - W \sin 5^O = ma_x \qquad \text{(a)}$$

The acceleration a_{x*} is: $a_{x*} = \Delta v / \Delta t = 20/40 = \tfrac{1}{2} \text{ m/s}^2$ (b)

Substituting Eq. (b) into Eq. (a) yields:

$$2F_f = m(\tfrac{1}{2} + g \sin 5^O) = 2{,}439 \text{ N} \qquad \text{(c)}$$

The maximum power output is:

$$P_{Out)Max} = 2F_f (v_{Max}) = 2{,}439 \,(20) = 48.78 \text{ kW} \qquad \text{(d)}$$

The maximum power input is:

$$P_{In)Max} = P_{Out)Max} / \mathcal{E} = 48.78/0.40 = 121.95 \text{ kW} \qquad \text{(e)}$$

The average power output is:

$$P_{Out)Ave} = 2F_f (v_{Ave}) = 2{,}439(10) = 24.39 \text{ kW} \qquad \text{(f)}$$

The average power input is:

$$P_{In)Ave} = P_{Out)Ave} / \mathcal{E} = 24.39/0.40 = 60.98 \text{ kW} \qquad \text{(g)}$$

7.5   POTENTIAL ENERGY

We have briefly discussed potential energy, when considering the work associated with changing the elevation y of a mass m or deforming a spring.

$$E_W = mg(y_1 - y_2) \qquad (7.6)$$

This work in Eq. (7.6) is considered as a negative quantity, because the particle lost potential energy as it moved from a higher elevation y_2 to a lower elevation y_1. (See Fig. 7.5).

Let's change the nomenclature to better reflect potential energy instead of work by rewriting Eq. (7.6) as:

$$E_P = E_W = mg(y_1 - y_2) = Wy \qquad\qquad (7.19)$$

where y is a height measurement from a specified datum plane, which is usually the Earth's surface.

When a particle moves upward in a gravitational force field the potential energy E_P increases, and when the particle moves downward E_P decreases. In most applications of Eq. (7.19), the gravitational acceleration g is treated as a constant; because y is very small compared to the radius of the Earth. However, for very large values of y, g is a variable dependent on y.

7.5.1 Stored Energy

Stored energy is often considered as a form of potential energy, because it may be available for use when required. This availability is similar to the potential energy due to the height of some mass that can be lowered to gain energy or raised to store energy when necessary.

In dynamics stored energy is usually restricted to springs that can be deformed, either compressed or tensioned, to store energy. The equation for potential energy stored in a spring is:

$$E_P = \tfrac{1}{2}\,kx^2 \qquad\qquad (7.20)$$

Energy may be stored in either compression or tension springs. Energy is stored in compression springs by applying an axial force that reduces its length. Conversely energy is stored in a tension spring by applying an axial force that increases its length.

7.5.2 Conservation of Energy

Energy can be converted from one form to another, but cannot be lost when measured globally. Energy is lost due to friction forces that oppose the direction of the motion of a particle. However, this energy is conserved because the energy lost due friction is converted to heat, which is included in the global energy accumulation in the Earth's atmosphere and in space.

Work energy can be converted to kinetic or potential energy. Potential energy can be converted to work energy or kinetic energy and kinetic energy can be converted to work energy or potential energy. We have previously written equations for the relationship between work, kinetic and potential energy as:

$$E_k)_1 + \Sigma E_W = E_k)_2 \qquad\qquad (7.13)$$

$$E_P = E_W = mg(y_1 - y_2) = Wy \qquad\qquad (7.19)$$

If we consider work where the forces are conservative, which implies that the work performed in moving a body from one point to another point is path independent. The work depends on only on the end points of the motion. However, the work performed by non-conservative forces depends of the path of the forces as they move from one point to another. Friction forces are non-conservative (or dissipative), because the work done in going from one point to another depends on the path followed.

If we assume the forces are conservative, the principle of conservation of energy is valid and it is apparent that:

$$E_k)_1 + E_P)_1 = E_k)_2 + E_P)_2 \qquad (7.21)$$

Equation (7.21) gives the combined kinetic and potential energy of a particle in an initial state and in a second state. This relation is useful in solving problems, where the particle changes elevation between the initial and final states.

EXAMPLE 7.9

A pendulum is fabricated with a 5 ft rod and a weight of 45 lb, as shown in Fig. E7.9. The pendulum is initially at rest. Determine the maximum velocity of the weight and the tension in the rod, when the weighs achieves its maximum velocity. Assume the weight of the rod is negligible compared to the weight attached to its end.

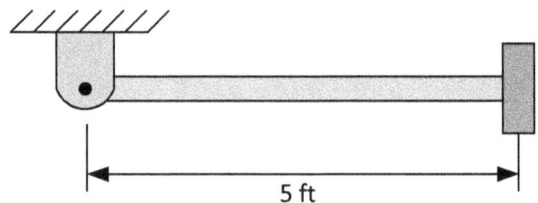

Fig. E7.9 A pendulum initially at rest in the 90 degree position.

Solution:

We will employ the principle of conservation of energy, write Eq. (7.21) and establish a datum plane 5 ft downward from the pin in the clevis:

$$E_k)_1 + E_P)_1 = E_k)_2 + E_P)_2 \qquad (7.21)$$

Because the pendulum is initially at rest $\qquad\qquad E_k)_1 = 0 \qquad$ (a)

Because the datum plane is 5 ft below the clevis pin: $\qquad E_P)_2 = 0 \qquad$ (b)

Then Eq. (7.21) reduces to: $\qquad\qquad E_P)_1 = E_k)_2 \qquad$ (c)

$$5\,W = \tfrac{1}{2}\,(W/g)v^2$$

$$v^2 = 10g = 321.7 \text{ ft}^2/\text{s}^2 \qquad \text{and} \qquad v = 17.94 \text{ ft/s} \qquad (d)$$

To determine the tension in the rod, prepare a free body diagram of the weight:

Write the equation of motion for the weight in the y direction, which coincides with the normal direction, when the weight is at the bottom of its swing.

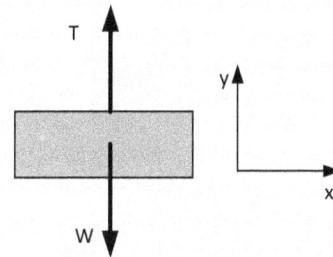

ΣF_y: $\qquad\qquad T - W = ma_n = mv^2/r \qquad\qquad$ (e)

$$T = W (1 + v^2/gr) = 45(1 + 2) = 135 \text{ lb} \qquad \text{(f)}$$

EXAMPLE 7.10

Two compression springs of equal length are nested together to form a shock limiter, as shown in Fig E7.10. A mass of 2 kg, initially at rest, is dropped from a height of 0.5 m, onto the top of the nested springs. The maximum deflection of the two springs under the impact of the weight is 0.2 m. Determine the spring rate of the inside spring, if the spring rate of the outside spring is 400 N/m.

Fig. E7.10 Photograph of nested helical coil springs. Courtesy of the Hoosier Spring Com.

Solution:

We will employ the relations for potential energy and equate the potential energy stored in the spring to the potential energy lost when the mass dropped compresses the two nested springs:

$$E_P)_{Stored} = \tfrac{1}{2} (k_{In} + k_{Out})(x)^2 = \tfrac{1}{2} (k_{In} + 400)(0.2)^2 = 0.02(k_{In} + 400) \qquad \text{(a)}$$

$$E_P)_{Stored} = 0.02 \, k_{In} + 8 \qquad \text{(b)}$$

The potential energy converted to stored potential energy due to the elevation change is:

$$E_P = Wy = 2g(0.5 + 0.2) = 13.73 \text{ N-m} \qquad \text{(c)}$$

Combining Eq. (b) and (c) gives:

$$k_{In} = (13.73 - 8)/0.02 = 286.5 \text{ N/m} \qquad \text{(d)}$$

EXAMPLE 7.11

A cylinder with a mass m= 4 kg is propelled up a smooth ramp with an initial velocity of 20 m/s, as shown in Fig. E7.11. The cylinder clears the ramp at point #2 and is in free flight until it lands at point #3. Determine the velocity of the cylinder at points #2 and #3 and the distance d it traveled in free flight.

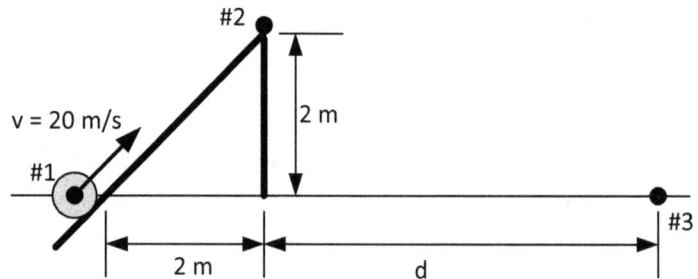

Fig. E7.11 Ramp geometry with cylinder in motion.

Solution:

We will present solutions to this Example using two different methods: first, by using the conservation of energy method as outlined above and second by using the equations of motion, as described in previous chapters.

Using the conservation of energy method:

Let's determine the kinetic and potential energy of the cylinder at the three different stations.

$$E_k)_1 = \tfrac{1}{2}\, mv_1{}^2 = \tfrac{1}{2}(4)(20)^2 = 800 \text{ N-m} \qquad \text{(a)}$$

$$E_P)_1 = Wy = 4g(0) = 0$$

$$E_k)_2 = \tfrac{1}{2}\, mv_2{}^2 = \tfrac{1}{2}(4)v_2{}^2 = 2v_2{}^2 \qquad \text{(b)}$$

$$E_P)_2 = Wy = mg(2) = 4g(2) = 8g \text{ N-m}$$

$$E_k)_3 = \tfrac{1}{2}\, mv_3{}^2 = \tfrac{1}{2}(4)v_3{}^2 = 2v_3{}^2 \qquad \text{(c)}$$

$$E_P)_3 = Wy = 4g(0) = 0$$

Consider the transition from point #1 to point #2:

$$E_k)_1 + E_P)_1 = E_k)_2 + E_P)_2 \qquad \text{(d)}$$

From Eqs. (a) and (b) we write:

$$800 + 0 = 2v_2{}^2 + 8g \qquad \text{or} \qquad v_2{}^2 = (800 - 8g)/2 = 360.8 \text{ m}^2/\text{s}^2$$

$$v_2 = 18.99 \text{ m/s} \qquad \text{(e)}$$

Consider the transition from point #1 to point #3:

$$E_k)_1 + E_P)_1 = E_k)_3 + E_P)_3 \qquad \text{(f)}$$

From Eqs. (a) and (c):

$$800 + 0 = 2v_3^2 + 0$$

$$v_3 = [400]^{1/2} = 20 \text{ m/s} \qquad \text{(g)}$$

The final velocity is the same as the initial velocity, because there is no change in elevation of the cylinder and no energy losses due to friction. To determine the distance d, consider the cylinder as a projectile in free flight when it leaves the ramp at position #2. Let's write the equation for the elevation y, as a function of time after the cylinder leaves the ramp.

$$y = y_0 + v_2 \sin \theta \, (t) - \tfrac{1}{2} g t^2 \qquad \text{(h)}$$

Note $\theta = 45^O$ and set Eq. (h) equal to zero, to give the quadratic equation for the time that the cylinder is in free flight:

$$y = 2 + (18.99)(0.7071)t - \tfrac{1}{2} (9.807) \, t^2 = 0$$

Rearranging terms gives:

$$4.9035 \, t^2 - 13.43 \, t - 2 = 0 \qquad \text{(i)}$$

Solving the quadratic equation yields the positive root as:

$$t = 2.878 \text{ s} \qquad \text{(j)}$$

The distance d traveled in free flight is:

$$d = v_2 \cos \theta \, (t) = 18.99(0.7071)(2.878) = 38.65 \text{ m} \qquad \text{(k)}$$

Using the equations of motion method:

Prepare a free body diagram of the cylinder at position #1, as shown in Fig E7.11:

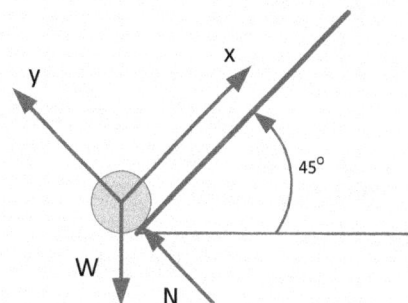

Fig. E7.11 The free body diagram of the cylinder at position #1.

Without friction (ramp is smooth), the cylinder slides up the ramp without rolling. We write $\Sigma F_x = ma_x$.

$$\Sigma F_x = ma_x = -W \cos 45^O = mg \cos 45^O = ma_x \qquad \text{(a)}$$

Reducing Eq. (a) give a_x as:

$$a_x = - g \cos 45^\circ = - 6.934 \text{ m/s}^2 \qquad \text{(b)}$$

Because the ramp is smooth, the acceleration is constant and we may write:

$$s = v_1 t + \tfrac{1}{2} a_x t^2 = [2^2 + 2^2]^{1/2} = 2.828 \text{ m} \qquad \text{(c)}$$

where the length of the ramp is s, and t is the time for the cylinder to move to position #2.

Substituting Eq. (b) into Eq. (c) and noting that a_x is a negative value gives a quadratic equation in t.

$$3.464 \, t^2 - 20 \, t + 2.828 = 0 \qquad \text{(d)}$$

The quadratic equation solver yields two roots for t as:

$$t = 0.1450 \text{ s and } 5.526 \text{ s} \qquad \text{(e)}$$

The first root (0.1450 s) is the correct solution, because the cylinder clears the ramp in a much shorter time than 5.526 s.

Knowing the time for the cylinder to move to position #2 enables us to compute the velocity v_2 as:

$$v_2 = v_1 + a_x t = 20 - 6.934(0.1450) = 18.99 \text{ m/s} \qquad \text{(f)}$$

This result is confirmed by the same value of the velocity v_2.

Next let's draw a free body diagram of the cylinder as it moves from Position #2 to Position #3.

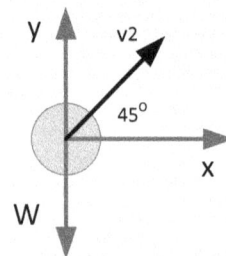

Fig. E7.11a The free body diagram of the cylinder in free flight with velocity v_2, as it leaves position #2 heading for position #3.

The equations of motion yield:

$$\Sigma F_x = ma_x = 0 \qquad \text{and} \qquad \Sigma F_y = -W = - mg = ma_y \qquad \text{(g)}$$

The second of Eq. (g) enables us to write:

$$a_y = - g = - 9.807 \text{ m/s}^2 \qquad \text{(h)}$$

Because a_y is a constant we can write:

$$y = y_0 + v_2 \sin 45^\circ t + \tfrac{1}{2} a_y t^2 = 0 \qquad \text{(i)}$$

Substituting numerical values into Eq. (i) and rearranging terms gives:

$$-4.903 \, t^2 + 13.44 \, t + 2.0 = 0 \qquad \text{(j)}$$

The quadratic equation solver yields two roots for the time of flight t. Selecting the positive root gives:

$$t = 2.882 \text{ s} \qquad \text{(k)}$$

Position #3 is determined from the velocity of the cylinder in the x direction multiplied by the time of flight.

$$x_3 = v_2 \cos 45^\circ t = 18.99(0.7071)(2.882) = 38.70 \text{ m}$$

This result confirms the position determined with the energy method.

EXAMPLE 7.12

A gun fabricated from a compression spring is shown in the figure to the right. The compression spring has a spring rate of k = 24 lb/in. The spring is initially compressed by 0.8 in. and then released to propel a ball with a weight of 4.0 oz. out of the barrel. Determine the exit velocity of the ball.

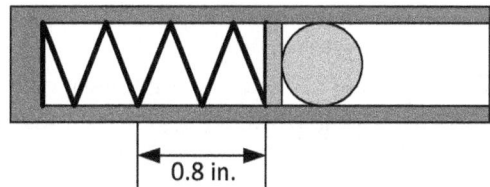

0.8 in.

Solution:

The energy stored in the spring becomes the initial potential energy:

$$E_P = \tfrac{1}{2} kx^2 = \tfrac{1}{2}(24)(0.8)^2 = 7.68 \text{ in-lb} \qquad \text{(a)}$$

The kinetic energy after the spring is released is given by:

$$E_k = \tfrac{1}{2} mv^2 = \tfrac{1}{2}(W/g) v^2 \qquad \text{(b)}$$

Equating the potential energy stored in the spring with the kinetic energy of the ball, as it exits the tube gives:

$$\tfrac{1}{2}(W/g) v^2 = 7.68 \qquad \text{(c)}$$

Note that g = 12(32.17) = 386.0 in/s^2 and 1.0 lb = 16 oz.

$$v^2 = 7.68(2)(32.17)(12)/(4/16) = 23{,}718 \text{ and } \qquad v = 154.0 \text{ in./s} = 12.83 \text{ ft/s} \qquad \text{(d)}$$

7.6   SUMMARY

The work E_W performed by a constant force F is:

$$E_W = F \int_{x_1}^{x_2} dx = F(x_1 - x_2) \qquad (7.1)$$

If F is not a constant but a function of x, the work performed is:

$$E_W = \int_{x_1}^{x_2} f(x)dx \qquad (7.2)$$

For a spring, with a spring rate k, the work to deform it is:

$$E_W = k \int_{x_1}^{x_2} xdx = (1/2)k\left[x_2^2 - x_1^2\right] \qquad (7.4)$$

Work is also performed by changing the elevation of a weight W:

$$E_W = mg(y_1 - y_2) = W(y_1 - y_2) \qquad (7.5)$$

The work due to the change in elevation of the particle in a gravitational f field is equal to the potential energy E_P of the particle.

$$E_W = W(y_1 - y_2) = E_P)_1 - E_P)_2 \qquad (7.6)$$

When the particle is moving along a curvilinear path, where the force F is a variable with r, the work E_W performed is given by:

$$E_W = \int_{r_1}^{r_2} F \bullet dr = \int_{s_1}^{s_2} F\cos\theta ds \qquad (7.8)$$

The equation of motion for the particle in the tangential direction gives:

$$F_t = ma_t = m\, v\, dv/dx \qquad (7.9)$$

The work performed is:

$$\int_{x_1}^{x_2} F_t dx = m\int_{v_1}^{v_2} vdv \qquad (7.10)$$

The kinetic energy is:

$$E_k = \tfrac{1}{2} m(v_2^2 - v_1^2) \qquad (7.12)$$

Energy is conserved, which enables us to write:

$$E_k)_1 + \Sigma E_W = E_k)_2 \qquad (7.13)$$

Power P is defined as the rate of change of work E_W with respect to time.

$$P = dE_W/dt \qquad (7.14)$$

Note that:

$$P = \mathbf{F} \bullet d\mathbf{r}/dt \qquad (7.15)$$

$$P = \mathbf{F} \bullet \mathbf{v} \qquad (7.16)$$

The power is a scalar quantity, because the dot product of two vectors yields a scalar result.

In the SI system of units, the power P is expressed in watts $\mathbf{W}$ where:

$$1\ \mathbf{W} = 1\ joule/s = 1\ J/s = 1\ N\text{-}m/s$$

In the U. S. Customary system of units, the power P is expressed in horsepower HP, where:

$$1\ \mathbf{HP} = 550\ ft\text{-}lb/s = 33{,}000\ ft\text{-}lb/min.$$

Efficiency is defined by:

$$\mathcal{E} = P_{Out}/P_{In} \qquad (7.17)$$

Efficiency may also be expressed in terms of energy E_{In} and energy E_{Out} of a system, if both are measured over the same time interval.

$$\mathcal{E} = E_{Out}/E_{In} \qquad (7.18)$$

Potential energy and work are equated giving:

$$E_P = E_W = mg(y_1 - y_2) = Wy \qquad (7.19)$$

The equation for potential energy stored in a spring is:

$$E_P = \tfrac{1}{2} kx^2 \qquad (7.20)$$

The principle of conservation of energy gives:

$$E_k)_1 + E_P)_1 = E_k)_2 + E_P)_2 \qquad (7.21)$$

CHAPTER 8

WORK AND ENERGY: RIGID BODIES

INTRODUCTION

In this chapter you will learn about work and the two types of energy — kinetic energy and potential energy E_P. You are already familiar with all of these forms of energy as they are related to particle motion. When dealing with rigid bodies determining the kinetic energy is more involved if the body is translating and rotating. You will establish the kinetic energy due to translation using the velocity of the centroid of the body and the kinetic energy due to rotation using the angular velocity of the body.

You will still deal with work energy, which is determined using essentially the same methods that were employed with particle motion; however, work energy is also developed due to moments rotating through an angle θ. The total work energy includes the part due to applied forces and the part due to moments.

The approach you will use with the determination of potential energy is also the same as you used with the change in elevation of a particle. However, with rigid bodies the elevation is measured with reference to the centroid of the rigid body. You learn to compute potential energy relative to a datum plane. When the position of a weight is above the datum plane the potential energy is positive. However, when the position of the weight is below the datum plane the potential energy is negative.

You will note that the work energy due the change in the length x of a spring in either tension or compression is the same as expressed previously. Energy is also stored in torsional springs that are twisted through some angle θ.

Recall that energy is conserved for particles and rigid bodies. If you correctly take all the forms of energy into account in the initial and final states of motion, a very useful relation is obtained that leads to the solution of many problems in a simple and direct manner.

CONCEPT PROBLEM

An amusement park ride, called Sky High Adventure, is illustrated in the figure below. A car with passengers weighing 1,200 lb is initially at rest at the top of the track H = 300 ft above ground level. The car descends down to ground level and enters a circular loop with a radius R = 100 ft. Determine the velocity of the car as it enters the circular loop and at the top of the loop. Also show the force holding the car to the track at the top of the loop. Model the car as a particle and neglect rail friction and drag.

An amusement park ride, with the car moving from a high elevation down to and around a circular loop.

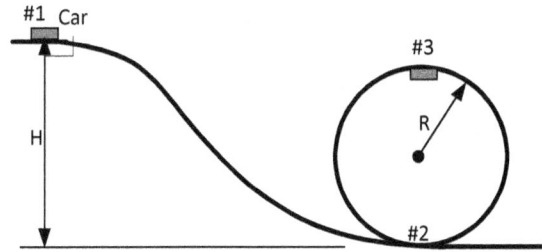

Solution:

The initial kinetic energy E_{k1} of the car is zero at position #1.

The initial potential energy of the car when it is located at position #1 is:

$$E_{P1} = Wy = 1,200(300) = 36 \times 10^4 \text{ ft-lb} \qquad (a)$$

The kinetic energy of the car at position #2 located at the base of the circular loop is:

$$E_{k2} = \tfrac{1}{2} mv_2^2 = \tfrac{1}{2} (1,200/32.17)v_2^2 = 18.65 \, v_2^2 \qquad (b)$$

$$E_{P2} = 0 \qquad \text{if we define the datum at position \#2 at the bottom of the loop.}$$

Then $\qquad E_{k2} = E_{P1} - E_{P2} = 18.65 \, v_2^2 = 36 \times 10^4 \qquad (c)$

$$v_2^2 = 36 \times 10^4 / 18.65 = 19,303 \text{ ft}^2/\text{s}^2 \qquad v_2 = 138.9 \text{ ft/s} \qquad (d)$$

At the position #3 located at the top of the loop the potential energy of the car is:

$$E_{P3} = Wy = 1,200(200) = 24 \times 10^4 \text{ ft-lb} \qquad (e)$$

And

$$E_{k3} = \tfrac{1}{2} mv_3^2 = \tfrac{1}{2} (1,200/32.17)v_3^2 = 18.65 \, v_3^2 \qquad (f)$$

Then $\qquad E_{k3} = E_{P1} - E_{P3} = (36 - 24) \times 10^4 = 18.65 \, v_3^2 = 12 \times 10^4 \qquad (g)$

$$v_3^2 = 12 \times 10^4 / 18.65 = 6,434 \text{ ft}^2/\text{s}^2 \qquad v_3 = 80.21 \text{ ft/s} \qquad (h)$$

The normal force acting on the car at position #3 due to its circular path at velocity is given by:

$$F_n = ma_n = (1,200/32.17)(v_3^2/r) \qquad (i)$$

$$F_n = (1,200/32.17)(6,434/100) = 2,400 \text{ lb} \qquad (j)$$

The force holding the car to the track at the top of the loop is calculated from summing the forces in the y direction on the free body diagram of the car at its top position in the circular loop, as shown below.

$\Sigma F_y :$ $\qquad\qquad F_n - W - N = 0$

$$N = F_n - W = 2400 - 1200 = 1{,}200 \text{ lb}$$

The centrifugal force F_n holds the car to the rails, with a sufficient (1,200 lb) margin.

8.1   KINETIC ENERGY

Consider a plane rigid body subjected to planar motion, as shown in Fig. 8.1. The body is moving in an inertial Oxy coordinate system, with its origin at point O. The origin at O is translating with a velocity v_O and the body is rotating with an angular velocity ω. We isolate a small mass dm in the body at point Q and identify its location with the position parameter **r**.

Fig. 8.1 A plane rigid body in planar motion.

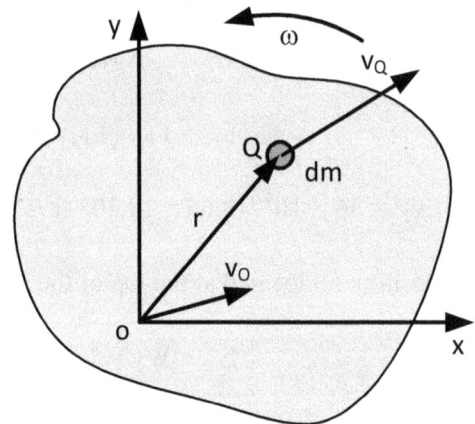

The kinetic energy dE_k of the incremental mass dm is:

$$dE_k = \tfrac{1}{2} \, dm \, v_Q^{\,2} \tag{8.1}$$

Integrating Eq. (8.1) yields:

$$E_k = \frac{1}{2}\int_m v_Q^{\,2} dm \tag{8.2}$$

Note that v_Q may be written as:

$$v_Q = v_O + v_{Q/O} \tag{8.3}$$

or

$$v_Q = (v_O)_x \, \mathbf{i} + (v_O)_y \, \mathbf{j} + \omega \, \mathbf{k} \times (x \, \mathbf{i} + y \, \mathbf{j})$$

$$v_Q = [(v_O)_x - \omega y] \, \mathbf{i} + [(v_O)_y + \omega x] \, \mathbf{j} \tag{8.4}$$

Square v_Q using the dot product to give:

$$v_Q^{\,2} = v_Q \bullet v_Q = [(v_O)_x - \omega y]^2 + [(v_O)_y + \omega x]^2$$

$$v_Q^2 = v_{O)x}^2 - 2v_{O)x}\,\omega y + \omega^2 y^2 + v_{O)y}^2 + 2v_{O)y}\,\omega x + \omega^2 x^2$$

$$v_Q^2 = v_O^2 - 2v_{O)x}\,\omega y + 2v_{O)y}^2\,\omega x + \omega^2 r^2 \tag{8.5}$$

Substituting Eq. (8.5) into Eq. (8.2) yields:

$$E_k = (1/2)v_O^2\int_m dm - v_{O)x}\,\omega\int_m y\,dm + v_{O)y}\,\omega\int_m x\,dm + (1/2)\omega^2\int_m r^2 dm$$

or

$$E_k = (1/2)mv_O^2 - v_{O)x}\,\omega\bar{y}m + v_{O)y}\,\omega\bar{x}m + (1/2)I_O\omega^2 \tag{8.6}$$

If the origin of the Oxy coordinate system is taken at the body's mass center G, then:

$$E_k = (1/2)mv_G^2 + (1/2)I_G\omega^2 \tag{8.7}$$

Equation (8.7) gives the kinetic energy associated with both translation and rotation. If we have translation without rotation $\omega = 0$ and Eq. (8.7) reduces to:

$$E_k = (1/2)mv_G^2 \tag{8.8}$$

When the origin of the Oxy coordinate system is taken at the body's mass center G, and if the body is rotating about point G without translation, then $v_G = 0$ and then Eq. (8.7) reduces to:

$$E_k = (1/2)I_G\omega^2 \tag{8.9}$$

However, if the fixed axis of rotation is say point O, then we write:

$$v_G = r_G\,\omega \tag{8.10}$$

and

$$E_k = (1/2)mr_G^2\omega^2 + (1/2)I_G\omega^2 = (1/2)(mr_G^2 + I_G)\omega^2$$

$$E_k = (1/2)I_O\omega^2 \tag{8.11}$$

Note we used the parallel axis theorem to show that $I_O = mr_G^2 + I_G$

The kinetic energy may also be written in terms of the body's angular velocity about its instantaneous center as:

$$E_k = (1/2)I_{IC}\omega^2 \tag{8.12}$$

8.2   WORK ENERGY DUE TO APPLIED FORCES

We have previously considered work energy due to forces applied to particles and rigid bodies. We will summarize the results obtained previously, because they are useful in this chapter.

Work energy due to a variable force acting on a rigid body is given by:

$$E_W = \int F \bullet dr = \int_s F \cos\theta\, ds \tag{8.13}$$

Work energy due to a constant force F acting on a rigid body:

$$E_W = (F \cos \theta)s \tag{8.14}$$

Work energy due the change in the elevation y of the weight W of a rigid body is given by:

$$E_W = Wy \tag{8.15}$$

Note that this expression can be equated to the potential energy of the rigid body.

Work energy due the change in the length x of a spring in either tension or compression is given by:

$$E_W = \tfrac{1}{2}\, k(x_2{}^2 - x_1{}^2) \tag{8.16a}$$

Work energy E_{WT} due the change in the angle of rotation of a torsion spring is given by:

$$E_{WT} = \tfrac{1}{2}\, k_T\, (\theta_2{}^2 - \theta_1{}^2) \tag{8.16b}$$

where k_T is the spring rate of the torsion spring.

8.3   WORK ENERGY DUE TO A MOMENT

Consider a moment M acting on a rigid planar body, as shown in Fig. 8.2. The moment is a pure couple, formed by two equal and opposite forces separated by some distance. The incremental work performed by the moment M is:

$$dE_W = M\, d\theta$$

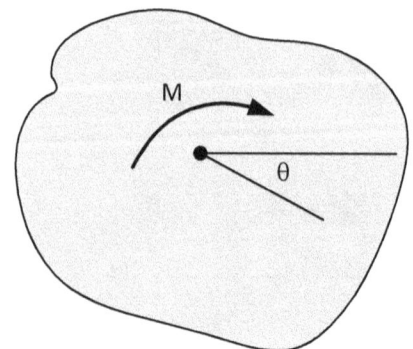

Fig. 8.2 A moment acting through an angle θ produces work energy.

Integrating yields:

$$E_W = \int_{\theta_1}^{\theta_2} M d\theta \qquad (8.17)$$

If the moment M is constant, Eq. (8.17) can be written as:

$$E_W = M(\theta_2 - \theta_1) \qquad (8.18)$$

EXAMPLE 8.1

A disk with a diameter of 18 in. and a weight of 45 lb is subjected to a force F = 14 lb, as shown in Fig. E8.1. If the disk is initially at rest, determine the angular velocity after is has rotated through ten revolutions.

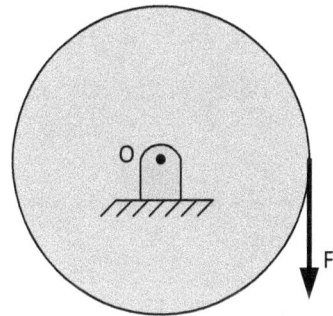

Fig. E8.1 A disk supported, by a bearing at point O, rotates due to the applied force F = 14 lb.

Solution:

The initial kinetic energy plus the work energy equals the final kinetic energy after ten revolutions. The initial kinetic energy is:

$$E_{k1} = 0 \qquad (a)$$

The force F travels through a distance s = r θ = (0.75) 2π (10) = 15π = 47.12 ft

The work due to the force F is:

$$E_W = F\,s = 14\,(47.12) = 659.7 \text{ ft-lb} \qquad (b)$$

$$E_{k1} + E_W = E_{k2} \qquad (c)$$

and

$$E_{k2} = \tfrac{1}{2} I_O \omega^2 = \tfrac{1}{2} (\tfrac{1}{2})(W/g)r^2 \omega^2 = \tfrac{1}{4} (45/32.17)(0.75)^2 \omega^2 = 0.1967 \omega^2 \qquad (d)$$

Substituting Eqs. (a), (b) and (d) into Eq. (c) yields:

$$\omega^2 = 659.7/0.1967 = 3{,}354 \text{ rad}^2/\text{s}^2$$

and

$$\omega = 57.91 \text{ rad/s} \qquad (e)$$

We could have determined the same result for the work performed by the force F by using the relation that $E_W = M\theta$:

Note that the moment M is given by $M = F \times r$ and the work $E_W = M\theta = F \times r \times \theta = F \times s$, which is the same as the relation shown between Eqs. (a) and (b) above. .

EXAMPLE 8.2

A rod 6 m in length with a mass m = 20 kg is subjected to the forces and moment, shown in Fig. E8.2. If the rod is initially at rest, determine its angular velocity ω, after it rotates through 6 revolutions.

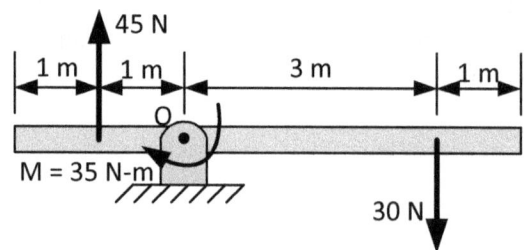

Fig. E8.2 A rod is subjected to two forces and a moment.

Solution:

Let's determine the work energy due to the two forces and the moment and then equate that energy to the final kinetic energy to calculate the angular velocity of the rod after six rotations.

Because the rod is initially at rest, the initial kinetic energy is:

$$E_{k1} = 0 \qquad (a)$$

The work energy is:

$$E_{W1} = F_1 \, s_1$$
$$E_{W2} = F_2 \, s_2 \qquad (b)$$
$$E_{W3} = M \, \theta$$

where s_1 is the distance through which the 45 N force travels and s_2 is the distance the 30 N force travels:

$$s_1 = 1(12\pi) = 12\pi$$
$$s_2 = 3(12\pi) = 36\pi \qquad (c)$$
$$\theta = 6(2\pi) = 12\pi$$

Note 12π is the angle generated by 6 revolutions of the rod.

$$E_{W1} = 45(12\pi) = 540\pi$$
$$E_{W2} = 30(36\pi) = 1,080\pi \qquad (d)$$
$$E_{W3} = 35(12\pi) = 420\pi$$

And

$$\Sigma E_W = 2040\pi \qquad (e)$$

$$E_{k2} = \tfrac{1}{2} I_O \, \omega^2 \qquad\qquad\qquad \text{(f)}$$

Note:
$$I_O = I_G + md^2 \qquad\qquad\qquad \text{(g)}$$

$$I_O = (1/12)mL^2 + md^2 = m(L^2/12 + d^2) \qquad\qquad \text{(h)}$$

$$I_O = m(L^2/12 + d^2) = 20[(6)^2/12 + (1)^2] = 80 \text{ kg-m}^2 \qquad \text{(i)}$$

The energy relation is:
$$E_{k1} + \Sigma E_W = E_{k2} = \tfrac{1}{2} I_O \, \omega^2 \qquad\qquad \text{(j)}$$

$$0 + \tfrac{1}{2} I_O \, \omega^2 = 40 \, \omega^2 = 2{,}040\pi \qquad\qquad \text{(k)}$$

$$\omega^2 = 160.2 \text{ rad}^2/\text{s}^2 \qquad\qquad \omega = 12.65 \text{ rad/s} \qquad \text{(m)}$$

EXAMPLE 8.3

A long thin rod with wheels at both ends is acted upon by a force F = 900 N, as shown in Fig. E8.3. The mass of the rod is 85 kg, its length L is 8 m, and it is initially at rest. A 900 N force drives the rod from its 45° orientation up the wall until it is vertical. Determine the angular velocity ω of the rod when it reaches the vertical position.

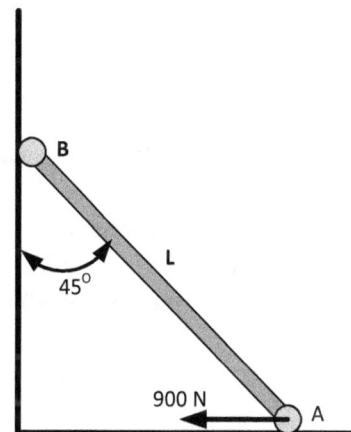

Fig. E8.3 A rod with wheels moves to the left and rotates until it is vertical.

Solution:

Let's prepare a drawing of the rod, as it rotates from its initial position to its final vertical orientation. We show an intermediate position of the rod in Fig. E8.3a, with the instantaneous center identified. When the rod moves to its final position the instantaneous center becomes located at point B

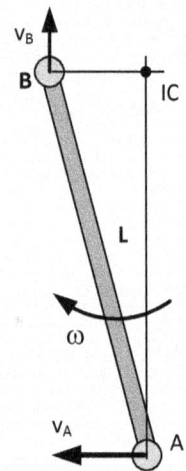

Fig. E8.3a The location of the instantaneous center (IC) at the position shown.

Fig 8.3b With the rod in its vertical position $v_A = 0$, because the rod is against the wall and its vertical movement ceases.

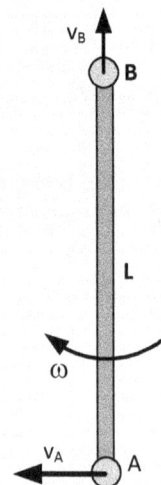

Determine the kinetic energies and the work energies.

The initial kinetic energy is:

$$E_{k1} = 0 \qquad \text{(a)}$$

The work performed by the 900 N force, as it moves to the left a distance L cos 45° is:

$$E_{W1} = 900 \, L \cos 45^O = 900(8)(0.7071) = 5,091 \text{ N-m} \qquad \text{(b)}$$

The work required to elevate the center of the mass of the rod, as it moved to the vertical position is:

$$E_{W2} = W(y_1 - y_2) = 85 \, (9.807)[4\sin 45^O - 4] = -976.6 \text{ N-m} \qquad \text{(c)}$$

$$\Sigma E_w = E_{W1} + E_{W2} = 5,091 - 976.6 = 4,114 \text{ N-m} \qquad \text{(d)}$$

The final kinetic energy is:

$$E_{k2} = \tfrac{1}{2} \, I_{IC} \, \omega^2 \qquad \text{(e)}$$

$$I_{IC} = 1/3mL^2 = (1/3)(85)(8)^2 = 1,813 \text{ N-m-s}^2 \qquad \text{(f)}$$

$$E_{k2} = \frac{1}{2}\,(1{,}813)\,\omega^2 = \Sigma E_w = 4{,}114 \text{ N-m} \qquad\qquad (g)$$

$$\omega^2 = 4{,}114/906.5 = 4.538 \qquad\qquad \omega = 2.130 \text{ rad/s} \qquad (h)$$

8.4   CONSERVATION OF ENERGY

In the previous chapter, we introduced the concept of conservation of energy for a particle. The conservation concept is the same when we are dealing with a rigid body. Energy is conserved providing that the forces involved are conservative, which implies that the energy involved is independent of the path of the motion. To illustrate this concept consider the potential energy of a weight W located at two different elevations, as shown in Fig. 8.3. We determine potential energy of a rigid body by its elevation, which is referenced to a horizontal datum plane.

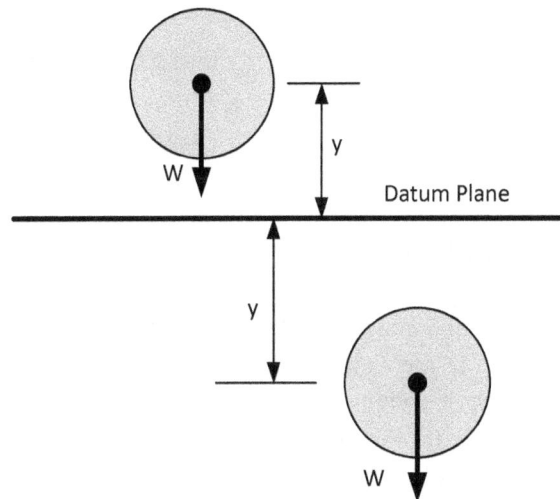

Fig. 8.3 A rigid body located above and below the datum plane.

The potential energy E_P of the rigid body in Fig. 8.3 that is above the datum plane a distance y is:

$$E_P = Wy_+ \qquad\qquad (8.19)$$

This potential energy is considered positive, because it can perform work because its elevation is reduced, when it moves toward the datum plane.

The potential energy E_P of the rigid body in Fig. 8.3 that is below the datum plane a distance y is:

$$E_P = -\,Wy_- \qquad\qquad (8.20)$$

This potential energy is considered negative, because we must perform work on the weight to move it against the gravitational field toward the datum plane.

Potential energy can be stored in a spring. When a body is attached to the spring, the elastic energy stored in the spring is transferred to the rigid body adding to its potential energy. On the other hand, the spring can extract energy from a rigid body, if the energy in the body is used to elongate a tension spring or compress a compression spring. We determine the energy stored in a spring by:

$$E_{PS} = \tfrac{1}{2}\,k(x_2{}^2 - x_1{}^2) \qquad\qquad (8.21)$$

The total potential energy of a spring and rigid body combined is due to both the elastic and the gravitational potential energy as indicated by:

$$\Sigma E_P = E_P + E_{PS} \qquad\qquad (8.22)$$

We must also consider work energy ΣE_W that can be added or removed from the system, as illustrated in Example 8.2.

Kinetic energy $E_k = \tfrac{1}{2}\,mv^2$ and/or $E_k = \tfrac{1}{2}\,I\omega^2$, due to translation and rotation in the initial and final state of motion, must also be considered when applying the principles of conservation of energy.

Taking all forms of the energy into account in the initial and final states of motion yields:

$$E_{k1} + \Sigma E_W + E_{P1} = E_{k2} + E_{P2} \qquad\qquad (8.23)$$

EXAMPLE 8.4

A bar with a mass of 60 kg is initially at rest in a horizontal orientation, as shown in Fig. E8.4. A spring, attached at the right end of the bar, is not extended or compressed in this initial state. When the bar is released, it swings downward to its vertical orientation and beyond. If the spring rate is 60 kN/m, determine the angular velocity ω when the bar swings to its vertical position.

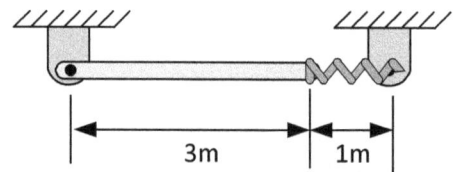

Fig. E8.4 The bar in its initial at rest position.

Solution:

Prepare a drawing of the bar in its vertical orientation.

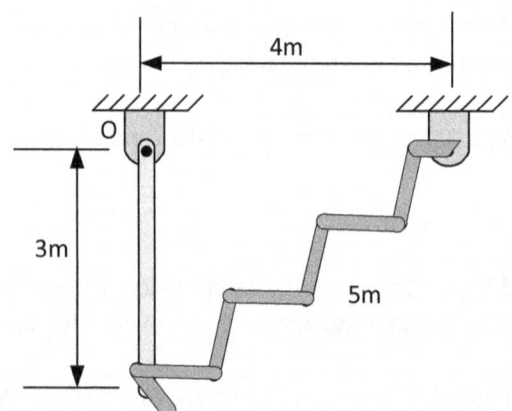

Fig. E8.4a The bar in its vertical position.

We note that the spring has extended from its initial free length of 1 m to the new length of 5 m. The work performed to stretch the spring E_{WS} is:

$$E_{WS} = \tfrac{1}{2}\,ks^2 = \tfrac{1}{2}(60)(5-1)^2 = 480 \text{ N-m} \tag{a}$$

The system gained potential energy, when the center of gravity of the bar rotated lower:

$$E_P = Wy = 60g(1.5) = 90(9.807) = 882.6 \text{ N-m} \tag{b}$$

The initial kinetic energy $E_{k1} = 0$. The final kinetic energy is:

$$E_{k2} = \tfrac{1}{2}\,I_O\,\omega^2 \tag{c}$$

where

$$I_O = 1/3mL^2 = 0.333(60)(3)^2 = 180 \text{ kg-m}^2$$

Then substituting into Eq. (c) we obtain:

$$E_{k2} = \tfrac{1}{2}(180)\,\omega^2 = 90\,\omega^2 = \tag{d}$$

Conservation of energy enables us to write:

$$E_{k1} + E_P - E_{WS} = E_{k2} \tag{e}$$

$$0 + 882.6 - 480 = 90\,\omega^2 \tag{f}$$

$$\omega^2 = 402.6/90 = 4.473 \text{ rad}^2/\text{s}^2 \qquad \omega = 2.115 \text{ rad/s} \tag{g}$$

EXAMPLE 8.5

Let's reconsider Example 8.4 with the bar and spring, as shown in Fig. E8.5. A bar with a mass of 60 kg is initially at rest in a horizontal orientation. A spring attached at the right end of the bar is not extended or compressed in this initial state. When the bar is released it swings downward extending the spring. We select a spring rate sufficiently large to limit the rotation of the bar to less than 90 degrees. If the spring rate is 200 kN/m, determine the angular position of the bar, when its angular velocity first becomes zero.

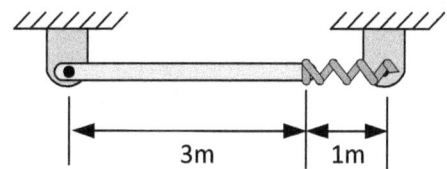

Fig. E8.5 The bar and spring in their initial position. The spring in its initial position is not extended or compressed.

3m   1m

Solution:

The bar is at rest in its initial and final states; hence, the kinetic energy is $E_{k1} = E_{k2} = 0$. The change in potential energy E_P equals the work E_{WS} required to extend the spring as the bar swings downward.

To better understand the geometry of the final state, let's draw the bar with $\omega = 0$, and the spring extended, as shown in Fig. E8.5a.

The change in potential energy E_P is:

$$E_P = Wy = mg(1.5 \sin \theta) = (1.5)(60)(9.807)\sin \theta = 882.6 \sin \theta \qquad (a)$$

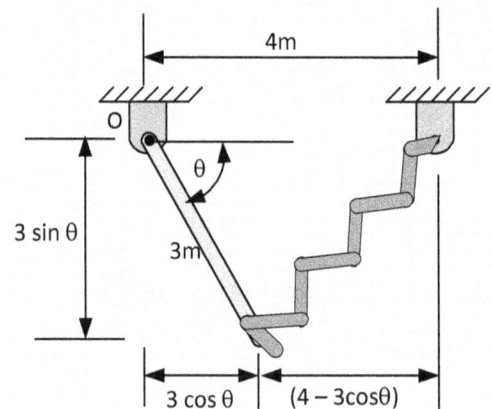

Fig. E8.5a Sketch showing spring in its extended position.

The final length L of the spring is:

$$L = [(3\sin \theta)^2 + (4 - 3\cos \theta)^2]^{1/2}$$

The amount the spring stretches is the final length minus the original free length of 1.0 m:

$$s = L - 1 = [(3\sin \theta)^2 + (4 - 3\cos \theta)^2]^{1/2} - 1 \qquad (b)$$

The energy to extend the spring is:

$$E_{WS} = \tfrac{1}{2} ks^2 = \tfrac{1}{2}(200)s^2 = 100\, s^2 \qquad (c)$$

Combine Eqs. (b) and (c) to obtain:

$$E_{WS} = 100\, s^2 = 100\{[(3\sin \theta)^2 + (4 - 3\cos \theta)^2]^{1/2} - 1\}^2 \qquad (d)$$

Use conservation of energy and equate E_P to E_{WS} to obtain:

$$E_{WS} = 100\{[(3\sin \theta)^2 + (4 - 3\cos \theta)^2]^{1/2} - 1\}^2 = E_P = 882.6 \sin \theta \qquad (e)$$

Use a spreadsheet to solve Eq. (e) numerically and obtain: $\theta = 64.3^O$

EXAMPLE 8.6

The disk, pinned at B with a 0.4 m diameter and a mass of 100 kg, is shown in Fig. E8.6. A cable attached to the rim of the disk supports a mass m_A = 90 kg. This mass A is moving downward with a velocity v_A = 2.2 m/s. Another mass m_C = 135 kg slides on a level surface with a coefficient of friction μ = 0.20. A spring with a spring rate of 70 N/m is attached by cable to the perimeter of the disk. Initially the spring is not compressed or extended. Determine the velocity v_A after block A has moved downward a distance of 1.2 m.

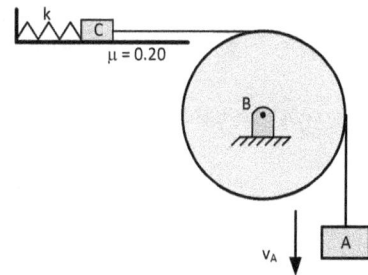

Fig. E8.6 The system includes a disk B, a weight A, a block C and a spring.

Solution:

We will employ conservation of energy in this solution. We will calculate the initial and final kinetic energy, the potential energy due to block, A and work energy to overcome friction and to extend the spring at block C.

Let's begin by drawing three free body diagrams: for block A, disk B and Block C.

Fig. E8.6a The FBD of the weight A with a mass of 90 kg and a velocity v_A .

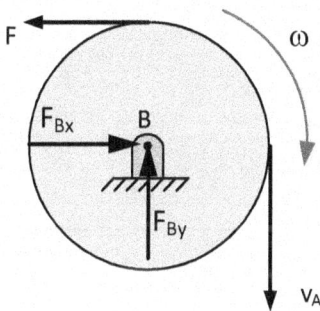

Fig. E8.6b The FBD of the disc B with a mass of 100 kg. The disk is rotating with an angular velocity of ω.

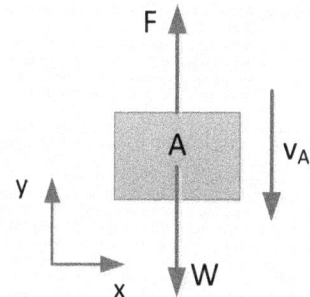

Fig. E8.6c The FBD of block C with a mass of 135 kg. Block C slides on a surface with a coefficient of friction of 0.2 and with a velocity of v_C.

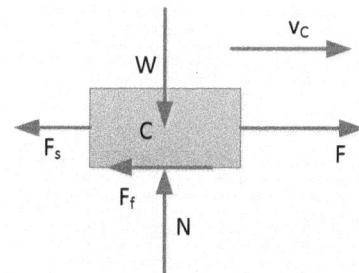

Begin by determining the initial kinetic energy E_{k1} of the entire system as:

$$E_{k1} = \tfrac{1}{2}\, m_A\, v_A^2 + \tfrac{1}{2}\, I_B\, \omega^2 + \tfrac{1}{2}\, m_C\, v_C^2 \qquad \text{(a)}$$

Note: $I_B = (1/2)m_B\, r_B^2$ $\omega = v_A/r_B$ $v_C = v_A$ (b)

From Eqs. (a) and (b) we obtain:

$$E_{k1} = \tfrac{1}{2}\, m_A\, v_A^2 + (1/4)\, m_B\, v_A^2 + \tfrac{1}{2}\, m_C\, v_A^2$$

Substituting numerical values into this equation yields:

$$E_{k1} = \tfrac{1}{2}\,(90)(2.2)^2 + \tfrac{1}{4}\,(100)(2.2)^2 + \tfrac{1}{2}\,(135)(2.2)^2 = 217.8 + 121 + 326.7 = 665.5 \text{ N-m} \qquad (c)$$

The work to extend the spring a distance of 1.2 m is:

$$E_{WS} = \tfrac{1}{2}\, ks^2 = \tfrac{1}{2}\,(70)(1.2)^2 = 50.4 \text{ N-m} \qquad (d)$$

Clearly N = W from Fig. E8.6c. The work to overcome friction generated by sliding block C is:

$$E_{Wf} = F_f\, s = N\mu s = 135(9.807)(0.2)(1.2) = 317.7 \text{ N-m} \qquad (e)$$

The potential energy E_P acquired due to reducing the elevation of m_A is:

$$E_P = Wy = 90(9.807)(1.2) = 1,059 \text{ N-m} \qquad (f)$$

The kinetic energy at the new position, after mass A moves 1.2 m is:

$$E_{k2} = \tfrac{1}{2}\, m_A\, v_{A2}^2 + \tfrac{1}{4}\, m_B\, v_{A2}^2 + \tfrac{1}{2}\, m_C\, v_{A2}^2 = (45 + 25 + 67.5)\, v_{A2}^2 = 137.5\, v_{A2}^2 \qquad (g)$$

From conservation of energy, we obtain:

$$E_{k2} = E_{k1} + E_P - E_{WS} - E_{Wf} = 665.5 + 1,059 - 50.4 - 317.7 = 1,356.4 \text{ N-m} \qquad (h)$$

Then
$$E_{k2} = 137.5\, v_A^2 = 1,356.4 \qquad (i)$$

$$v_{A2}^2 = 1,356.4/137.5 = 9.865 \qquad\qquad v_{A2} = 3.141 \text{ m/s} \qquad (j)$$

$$\omega_{B2} = v_{A2}/r_B = 3.141/0.2 = 15.70 \text{ rad/s} \qquad (k)$$

EXAMPLE 8.7

Two spur gears are illustrated in Fig. E8.7. Gear B with a mass of 6 kg drives gear A that has a mass of 20 kg. The radius of gyration of gear A is 0.4 m and the radius of gyration of gear B is 0.16 m. The gears are initially at rest. The motor driving gear B provides a moment M = 12 N-m. Determine the number of revolutions gear B turns before it achieves an angular velocity of 900 RPM. Also calculate the force acting on the gear teeth of gear A, when gear B achieves the 900 RPM angular velocity.

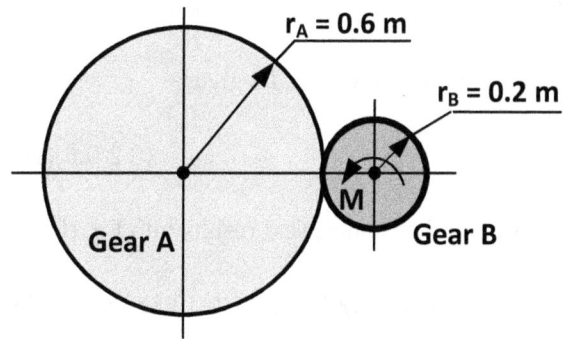

Fig. E8.7 Gear B, the pinion gear, drives gear A.

$r_A = 0.6$ m

$r_B = 0.2$ m

Solution:

We will employ energy methods to determine the number of revolutions of gear B, before it achieves the specified angular velocity.

Let's begin by drawing free body diagrams of the two gears:

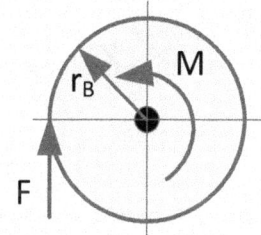

Fig. E8.7a The FBD of the pinion gear (gear B). We assume the forces and moments at the bearings do not perform work; hence, they are not shown here.

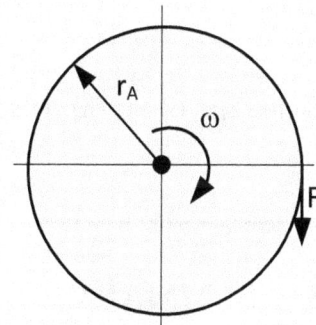

Fig. E8.7b The FBD of the driven gear (gear A). We assume the forces and moments at the bearings do not perform work; hence, they are not shown here.

Because the gears are initially at rest, their initial kinetic energy is:

$$E_{kA1} = E_{kB1} = 0 \qquad \text{(a)}$$

The kinetic energy of the system, when gear B achieves 600 RPM is:

$$E_{k2} = \tfrac{1}{2}\, I_A\, \omega_A^2 + \tfrac{1}{2}\, I_B\, \omega_B^2 \qquad \text{(b)}$$

The mass moments of inertia are:

$$I_A = m_A\, k_A^2 = 20(0.4)^2 = 3.2 \text{ kg-m}^2$$

$$I_B = m_B\, k_B^2 = 6(0.16)^2 = 0.1536 \text{ kg-m}^2 \qquad \text{(c)}$$

Note that: $\qquad r_A \, \omega_A = r_B \, \omega_B \qquad$ or $\qquad \omega_A = (r_B / r_A) \, \omega_B \qquad$ (d)

Converting 900 RPM to rad/s gives: $\qquad \omega_B = 94.25$ rad/s $\qquad$ (e)

Then $\qquad \omega_A = (0.2/0.6)(94.25) = 31.42$ rad/s $\qquad$ (f)

Substituting these numerical results into Eq. (b) yields:

$$E_{k2} = \tfrac{1}{2}\,(3.2)(31.42)^2 + \tfrac{1}{2}\,(0.1536)(94.25)^2 = 1{,}580 + 682.2 = 2{,}262 \text{ N-m} \qquad (g)$$

The work performed by the moment M is:

$$E_W = M\,\theta_B = 12\,\theta_B \qquad (h)$$

Note that $E_W = E_{k2}$, then from Eqs. (g) and (h), we write:

$$\theta_B = 2{,}262/12 = 188.5 \text{ rad} \qquad (i)$$

To determine the force acting on gear A, we note the kinetic energy of gear A is:

$$E_{k2})_{\text{Gear A}} = \tfrac{1}{2}\,I_A\,\omega_A^{\,2} = \tfrac{1}{2}\,(3.2)(31.42)^2 = 1{,}580 \text{ N-m} \qquad (j)$$

The work performed by this force on gear A is:

$$E_W = Fs = F\,\theta_A\,r_A = F\,\theta_B\,r_B = F\,(188.5)(0.2) = 37.7\,F \qquad (k)$$

Equating Eqs. (j) and (h) yields

$$F = 1{,}580/37.7 = 41.91 \text{ N} \qquad (l)$$

8.5 SUMMARY

The kinetic energy dE_k of an incremental mass dm is:

$$dE_k = \tfrac{1}{2}\,dm\,v_Q^{\,2} \qquad (8.1)$$

Integrating yields:

$$E_k = 1/2 \int_m v_Q^{\,2} dm \qquad (8.2)$$

The velocity v_Q is:

$$v_Q = v_O + v_{Q/O} \qquad (8.3)$$

$$v_Q = (v_O)_x\,\mathbf{i} + (v_O)_y\,\mathbf{j} + \omega\,\mathbf{k} \times (x\,\mathbf{i} + y\,\mathbf{j})$$

$$v_Q = [(v_O)_x - \omega y]\,\mathbf{i} + [(v_O)_y + \omega x]\,\mathbf{j} \qquad (8.4)$$

$$v_Q^2 = v_O^2 - 2v_{O)x}\,\omega y + 2v_{O)y}^2\,\omega x + \omega^2 r^2 \tag{8.5}$$

$$E_k = (1/2)v_O^2 - v_{O)x}\,\omega \bar{y}m + v_{O)y}\,\omega \bar{x}m + (1/2)I_O\omega^2 \tag{8.6}$$

If the origin of the Oxy coordinate system is taken at the mass center G of the rigid body, then:

$$E_k = (1/2)v_G^2 + (1/2)I_G\omega^2 \tag{8.7}$$

With fixed axis rotation about point O:

$$v_G = r_G\,\omega \tag{8.10}$$

$$E_k = (1/2)(mr_G^2 + I_G)\omega^2$$

$$E_k = (1/2)I_O\omega^2 \tag{8.11}$$

The kinetic energy in terms of the body's angular velocity about its instantaneous center is:

$$E_k = (1/2)I_{IC}\omega^2 \tag{8.12}$$

Work energy due to a variable force acting on a rigid body is:

$$E_W = \int F \cdot dr = \int_s F\cos\theta\,ds \tag{8.13}$$

Work energy due to a constant force F is:

$$E_W = (F\cos\theta)s \tag{8.14}$$

Work energy due the change in the elevation y of the weight W of a rigid body is:

$$E_W = Wy = E_P \tag{8.15}$$

Work energy due the change in the length x of a spring in either tension or compression is:

$$E_W = \tfrac{1}{2}\,k(x_2^2 - x_1^2) \tag{8.16a}$$

Work energy E_{WT} due the change in the angle of rotation of a torsion spring is given by:

$$E_{WT} = \tfrac{1}{2}\,k_T(\theta_2^2 - \theta_1^2) \tag{8.16b}$$

Work energy is also developed due to moments rotating through an angle θ that is determined from:

$$dE_W = M\,d\theta$$

Integrating yields:

$$E_W = \int_{\theta_1}^{\theta_2} M d\theta \tag{8.17}$$

If the moment M is a constant:

$$E_W = M(\theta_2 - \theta_1) \tag{8.18}$$

Potential energy is computed relative to a datum plane. When the position of the weight y is above the datum plane, the potential energy is positive and given by:

$$E_P = W y_+ \tag{8.19}$$

When the position of the weight is below the datum plane, the potential energy is negative and given by:

$$E_P = - W y_- \tag{8.20}$$

Potential energy E_{PS} can be stored by the elastic deformation of a spring and is determined from:

$$E_{PS} = \tfrac{1}{2} k(x_2^2 - x_1^2) \tag{8.21}$$

The total potential energy is due to both the elastic and the gravitational potential energy and is given by:

$$\Sigma E_P = E_P + E_{PS} \tag{8.22}$$

Taking all the forms of energy into account in the initial and final states of motion yields:

$$E_{k1} + \Sigma E_W + E_{P1} = E_{k2} + E_{P2} \tag{8.23}$$

CHAPTER 9

MOMENTUM AND IMPULSE: PARTICLES and RIGID BODIES

INTRODUCTION

In this chapter, you will learn about momentum and impulse for both particles and rigid bodies. Recall from previous chapters that linear momentum **M** of a particle is m**v**. Also recall that impulse is related to the change in the momentum. You will be introduced to angular momentum and learn how it is related to the angular velocity of a rigid body. Finally, the concept of angular impulse and its relationship to the change in angular momentum is introduced.

You will study the impact of spheres and rigid bodies and learn about the coefficient of restitution. Important in the study of impact is that momentum is conserved. With oblique impact you will follow the four rules listed below in solving this type of problem.

1. For unconstrained particles, the momentum of the system is conserved along the line of impact (x axis).

2. Momentum of sphere A is conserved along the plane of contact (y axis).

3. Momentum of sphere B is conserved along the plane of contact (y axis).

4. The coefficient of restitution is applied to the velocity components of spheres A and B relative to the line of impact.

CONCEPT PROBLEM

A box that is 0.6 m high and 0.4 m wide is travelling on conveyer belt A with a velocity v_1. The box strikes a stop B that is located between the belts A and C, as shown in the figure below: Determine the minimum velocity for the box to tip over at stop B and land on the conveyer belt C. The impact that occurs at stop B is plastic, with no rebound.

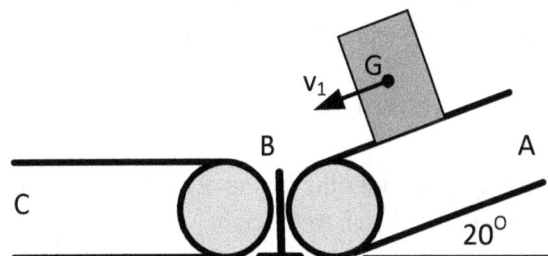

Transfer and tipping of the box between conveyer belts A and C.

Solution:

In this example, we will use the principal of conservation of angular momentum and the conservation of energy in the solution. We will also recognize that the box is essentially stationary at the tipping point associated with the minimum velocity.

Three sketches, shown below, illustrate the quantities in the conservation of momentum and energy relations that you will use in solving this problem.

The first sketch, found below, shows the box as it strikes the stop at point B. The left side of the graphic equation shows the box as it engages to stop. The right shows the box after it has tipped and begins to rotate.

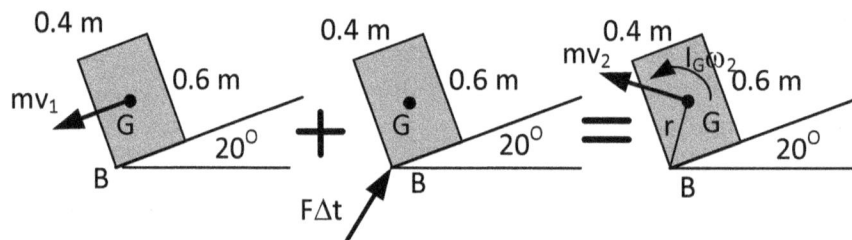

Let's use Eqs. (9.5) and (9.6) to write the relation for the conservation of angular momentum for the carton as it engages the stop at point B.

$$0.3 \, mv_1 + 0 = r \, mv_2 + I_G \, \omega_2 \tag{a}$$

$$r = \tfrac{1}{2} \, [(0.4)^2 + (0.6)^2]^{1/2} = 0.3606 \text{ m} \tag{b}$$

$$I_G = (1/12)m \, [(0.4)^2 + (0.6)^2] = 0.0433 \text{ m kg-m}^2 \tag{c}$$

Substituting Eqs. (b) and (c) into Eq. (a) gives:

$$0.3 \, mv_1 + 0 = 0.3606 \, mv_2 + 0.0433 \, m\omega_2 \tag{d}$$

Recall from previous chapters that:

$$v_2 = r\omega_2 \tag{e}$$

From Eqs. (d) and (e), you can solve for v_1 in terms of ω_2 as:

$$0.3v_1 = (0.3606)^2 \, \omega_2 + 0.0433 \, \omega_2 \qquad \text{and} \qquad v_1 = 0.5778 \, \omega_2 \tag{f}$$

Next let's consider the box as it is the process of engaging the stop at point B and just before tipping, as shown in the sketch to the right.

The positon of the box as it engages the stop B, and before it tips.

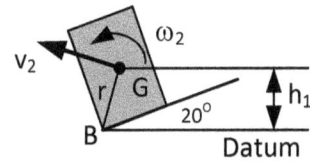

Next, let's consider the energy for the box engages the stop at point B.

Recall that the potential energy of the box, as it engages the stop at B is:

$$E_{P1} = Wh_1 = W\,r\,\sin(56.31^O + 20^O) = W(0.3606)\sin 76.31^O = 0.3504\,W \qquad (g)$$

The kinetic energy associated with the box, as it engages the step at B is:

$$E_{k1} = \tfrac{1}{2}\,mv_2^2 + \tfrac{1}{2}\,I_G\,\omega_2^2 \qquad (h)$$

$$E_{k1} = \tfrac{1}{2}\,m\,(r\omega_2)^2 + \tfrac{1}{2}(0.0433)\,m\,\omega_2^2 = \tfrac{1}{2}\,(W/g)(0.1733)\omega_2^2 \qquad (i)$$

The total energy as it engages the stop at B is:

$$E_{P1} + E_{k1} = 0.3504\,W + \tfrac{1}{2}\,(W/g)\,(0.1733)\omega_2^2 \qquad (j)$$

Next, let's examine the box when it is exactly at its tipping point. At the tipping point, the box is standing on its corner, with its diagonal vertical as shown below:

The box at its tipping point.

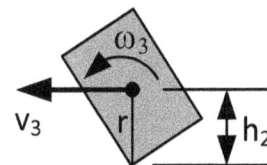

Because of the minimum input velocity requirement ($v_1 = 0$) for the box to tip, it is evident that the box is stationary at this instant and $v_3 = \omega_3 = 0$ and $E_{k2} = 0$.

The potential energy at the tipping point is:

$$E_{P2} = W\,h_2 = W\,r = 0.3606\,W \qquad (k)$$

where $r = 0.3606$ m is ½ the diagonal of the box.

Conservation of energy between these two positions gives:

$$E_{P1} + E_{k1} = E_{P2} + 0 \qquad (l)$$

$$E_{P1} + E_{k1} = 0.3504\,W + \tfrac{1}{2}\,(W/g)\,(0.1733)\omega_2^2 = 0.3606\,W \qquad (m)$$

$$\omega_2^2 = 2(0.3606 - 0.3504)g/(0.1733) = 0.1177g \qquad (n)$$

$$\omega_2{}^2 = 1.154 \text{ rad}^2/\text{s}^2 \qquad\qquad \omega_2 = 1.074 \text{ rad/s} \qquad\qquad \text{(o)}$$

$$v_1 = 0.5778 \, \omega_2 = 0.5778(1.074) = 0.6208 \text{ m/s} \qquad\qquad \text{(p)}$$

By determining the potential and kinetic energy of the box at three different positions during this dynamic event and by using the conservation of energy principle, we were able to calculate the linear and angular velocities of the box, as it engaged stop B and moved from one conveyer belt to the next one.

9.1   MOMENTUM AND IMPULSE OF A PARTICLE

Let's initially consider a particle with a mass m moving along a curvilinear path, with a velocity **v**, as shown in Fig. 9.1.

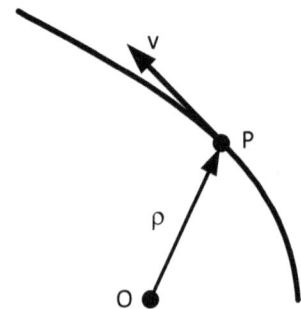

Fig. 9.1 A particle with a mass m is moving along a curvilinear path with a tangential velocity v.

We define the momentum **M** of the particle as:

$$\mathbf{M} = m\mathbf{v} \qquad\qquad \text{(9.1)}$$

Recall Newton's second law that gives the relation between the forces applied to any object and its momentum as:

$$\Sigma \mathbf{F} = d(m\mathbf{v})/dt = d\mathbf{M}/dt \qquad\qquad \text{(9.2)}$$

Integration of Eq. (9.2) over the time that the forces act yields:

$$\int_{t_1}^{t_2} \Sigma \mathbf{F} dt = \int_{M_1}^{M_2} d\mathbf{M} \qquad\qquad \text{(9.3)}$$

The impulse **I** of the particle is defined as:

$$\mathbf{I} = \int_{t_1}^{t_2} \Sigma \mathbf{F} dt \qquad\qquad \text{(9.4)}$$

Note that impulse is equal to the change in momentum that is given by:

$$\mathbf{I} = \Delta \mathbf{M} = \mathbf{M_2} - \mathbf{M_1} \qquad\qquad \text{(9.4a)}$$

9.1.1 Angular Momentum of a Particle

Angular momentum H_O is defined as the moment of the momentum about some point O. Examine Fig. 9.1 and note the location of point O and quantities ρ and v. The angular momentum H_O is given by:

$$H_O = \rho \times M = \rho \times (mv) \qquad (9.5)$$

where v is the tangential velocity of the particle.

A particle has mass without size; hence, the angular momentum about its center of gravity is zero. However, the particle has an angular momentum about any other point in space, as indicated in Eq. (9.5).

EXAMPLE 9.1

A particle, with a mass of 3.5 kg, is in motion along a curvilinear path and subjected to a constant force F. The particle initially is moving with a tangential velocity of 9 m/s. In the next 24 seconds, its velocity increases to 27 m/s. Determine the force acting on the particle.

Solution:

We will employ the relation between impulse and the change in linear momentum to determine the force acting on the particle. To begin, let's determine the initial and final values of the particle's momentum from Eq. (9.1).

$$M_1 = mv_1 = 3.5(9) = 31.5 \text{ kg-m/s}$$

$$\text{(a)}$$

$$M_2 = mv_2 = 3.5(27) = 94.5 \text{ kg-m/s}$$

From Eq. (9.4) we calculate the impulse **I** as:

$$I = M_2 - M_1 = 94.5 - 31.5 = 63 \text{ N-s} \qquad \text{(b)}$$

For a constant force, the impulse is determined from Eq. (9.3) as:

$$I = F (t_2 - t_1) = 24 \, F = 63 \text{ N-s} \qquad \text{(c)}$$

and

$$F = 63/24 = 2.625 \text{ N} \qquad \text{(d)}$$

9.2   MOMENTUM OF A RIGID BODY

9.2.1 Translation

When a rigid body is translating, the momentum is given by:

$$\mathbf{M} = mv_G \tag{9.6}$$

where $\mathbf{v}_G$ is the velocity of the center of mass of the rigid body, as shown in Fig. 9.2.

The angular momentum $\mathbf{H}_G$ relative to the center of gravity of a rigid body in translation is:

$$\mathbf{H}_G = I_G \, \boldsymbol{\omega} = 0 \tag{9.7}$$

Because $\omega = 0$ when the body is translating without rotation.

Fig. 9.2 A rigid body is translating in the y direction with the center mass moving at a velocity v_G.

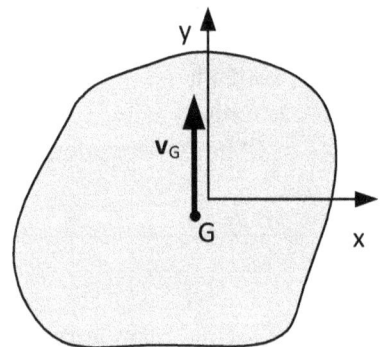

9.2.2 Rotation about a Fixed Axis

Consider a rigid body that is pinned at a fixed point O and rotating about that point with a an angular velocity ω, as shown in Fig. 9.3.

Fig. 9.3 A rigid body rotating about a fixed point with an angular velocity ω.

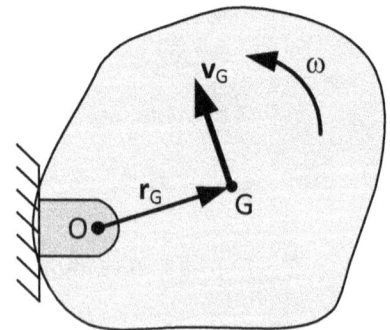

The linear momentum is given by:

$$\mathbf{M} = m \, \mathbf{v}_G = m \, r_G \, \boldsymbol{\omega} \tag{9.8}$$

Because O is the point about which the body is rotating, the angular momentum is given by:

$$\mathbf{H}_O = I_O \, \boldsymbol{\omega} \tag{9.9}$$

Using the parallel axis theorem, we know: $$I_O = I_G + m \, r_G^2 \tag{9.10}$$

9.2.3 General Plane Motion with Translation and Rotation

General plane motion involves both translation and rotation. Consider the rigid body, shown in Fig. 9.4, which is translating and rotating about the point O. We write the relation for the linear momentum of the body as:

$$\mathbf{M} = m \, \mathbf{v}_G \qquad (9.11)$$

The angular momentum relative to the center of gravity is:

$$\mathbf{H}_G = I_G \, \omega \qquad (9.12)$$

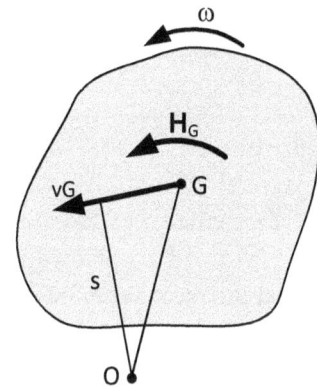

Fig. 9.4 A rigid body translating with a velocity v_G and rotating about point O with an angular velocity ω.

The angular momentum relative to a center of rotation O is:

$$\mathbf{H}_O = I_G \, \omega + s \, m \, \mathbf{v}_G \qquad (9.13)$$

where s is the length of the line perpendicular to the velocity vector $\mathbf{v}_G$, as shown in Fig. 9.4.

If point O is the instantaneous center IC:

$$\mathbf{H}_{IC} = I_{IC} \, \omega \qquad (9.14)$$

Let's consider an example, to demonstrate the use of these equations in calculating the angular momentum of a rigid body in general plane motion.

EXAMPLE 9.2

A rod with a mass of 20 kg and a length L = 8 m leans against a vertical wall at point B, as shown in Fig. E9.2. If the rod at point B, is moving downward with a velocity v_B = 1.2 m/s, determine the angular velocity ω of the rod, its velocity at point A, the velocity at its center of gravity, and the angular momentum $\mathbf{H}_{IC}$ about its instantaneous center.

Fig. E9.2 A rod is sliding down the vertical wall.

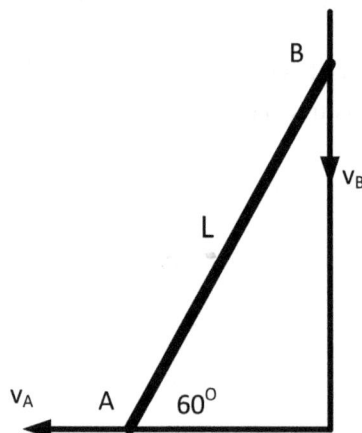

Solution:

Let's visualize the ladder as it slides down the wall and draw a diagram showing the velocity vectors at points A and B, as shown in Fig. E9.2a. Next draw perpendicular lines from the velocity vectors v_A and v_B that intersect to define the instantaneous center IC.

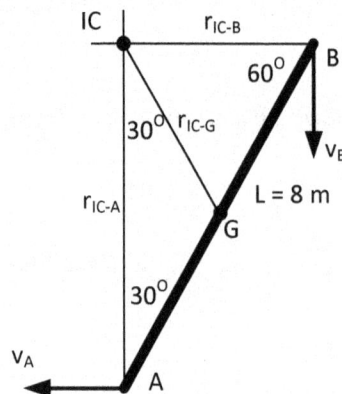

Fig. E9.2a A drawing showing the instantaneous center IC and several radii.

In Fig. E9.2a, it is clear that triangle IC-B-G is an equilateral triangle, with angles of 60°. It is also evident that triangle IC-G-A is an isosceles triangle with two 30° angles, as shown in Fig. E9.2a. From the geometry of these two triangles, it is clear that:

$$r_{IC-B} = 8 \sin 30^O = 4 \text{ m}$$

From the law of sines   (a)

$$r_{IC-G} = L/2 = 8/2 = 4 \text{ m}$$

We use the relation $v = r\omega$ to determine the angular velocity of the rod as:

$$\omega = v_B/r_{IC-B} = 1.2/4 = 0.3 \text{ rad/s}$$ (b)

The velocity of point A is:

$$v_A = r_{IC-A}\, \omega = 8 \cos 30^O (3) = 2.078 \text{ m/s}$$ (c)

The velocity of the center of gravity is:

$$v_G = r_{IC\text{-}G}\, \omega = 4(0.3) = 1.20 \text{ m/s} \qquad (d)$$

The angular momentum about the IC is given by Eq. (9.16) as:

$$\mathbf{H}_{IC} = I_{IC}\, \boldsymbol{\omega} \qquad (e)$$

Next we employ the parallel axis theorem to determine I_{IC} as:

$$I_{IC} = (1/12)\, mL^2 + m\, r_{IC\text{-}G}^2 = (1/12)(20)(8)^2 + 20(4)^2 = 106.7 + 320 = 426.7 \text{ kg-m}^2$$

$$\mathbf{H}_{IC} = I_{IC}\, \boldsymbol{\omega} = 426.7(0.3) = 128.0 \text{ kg-m}^2/\text{s} \qquad (f)$$

9.3 IMPULSE AND MOMENTUM OF A RIGID BODY

In section (9.1), we considered a particle moving with a tangential velocity v along a curvilinear path and introduced the concept of momentum. We also introduced the relationship between impulse and momentum in Eq. (9.4). The equations presented in Section 9.1, also apply to planar motion of rigid bodies, if we substitute $\mathbf{v}_G$ for $\mathbf{v}$ in the Equations (9.1), (9.2), (9.3) and (9.4a).

We have considered angular momentum in Section 9.2 and have developed the equation to determine it for translation, fixed axis rotation and general planar motion. However, we have not developed the relations between angular momentum and impulse for a rigid body.

9.3.1 Angular Impulse and Momentum

When considering planar motion of rigid bodies, we must account for angular momentum and impulse, because rigid bodies rotate in addition to translating. Rigid body rotation involves moments that are summed about the centroid G, which in turn produce angular acceleration given by:

$$\Sigma M_G = I_G\, \alpha = I_G\, (d\omega/dt) \qquad (9.15)$$

Multiplying through both sides of Eq. (9.15) by dt and integrating yields:

$$\Sigma \int_{t_1}^{t_2} M_G\, dt = I_G \int_{\omega_1}^{\omega_2} d\omega = I_G(\omega_2 - \omega_1) \qquad (9.16)$$

If we restrict the motion of the rigid body to rotation about a fixed axis located at some point O, then Eq. (9.16) can be expressed as:

$$\Sigma \int_{t_1}^{t_2} M_O\, dt = I_O(\omega_2 - \omega_1) \qquad (9.17)$$

It is interesting to compare a few equations to better understand the relationship between impulse and momentum. For linear momentum, we derived:

$$I = \int_{t_1}^{t_2} \sum F dt = \mathbf{M_2} - \mathbf{M_1} \qquad (9.4)$$

For angular momentum we derived:

$$\sum \int_{t_1}^{t_2} M_G dt = I_G \int_{\omega_1}^{\omega_2} d\omega = I_G(\omega_2 - \omega_1) \qquad (9.16)$$

Recall Eq. (9.12) and it is apparent that the angular impulse **IA** can be expressed as:

$$IA = \sum \int_{t_1}^{t_2} M_G dt = I_G(\omega_2 - \omega_1) = \mathbf{H_G})_2 - \mathbf{H_G})_1 \qquad (9.18)$$

The term on the left side of Eq. (9.18) is the angular impulse, and the terms on the far right hand side give the change in the angular momentum.

EXAMPLE 9.3

A double disk, shown in Fig E9.3, has an inner radius of 0.5 m and an outer radius of 1.0 m, with a mass of 125 kg and a radius of gyration about its center at point G of $k_G = 0.45$ m. The disk is initially at rest and then a constant force F = 120 N is applied. The disk rolls on the horizontal surface without slipping. Determine the linear velocity v_G and the angular velocity ω of the disk, after the force F has been applied for 5 s.

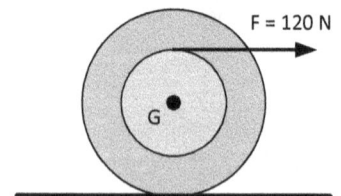

Fig. E9.3 A double disk rolls on the horizontal surface without slipping.

Solution:

Let's begin by preparing a free body diagram of the disk. By noting that the disc rolls without slipping, we recognize the point of contact with the horizontal surface is an instantaneous center IC.

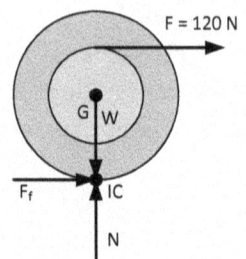

The disk is initially at rest; hence, the values of the initial linear and angular momentum are zero.

Let's consider a modified form of Eq. (9.18), where we replace the subscript G with IC.

$$\Sigma \int_{t_1}^{t_2} M_{IC} dt = I_{IC}(\omega_2 - \omega_1) = \mathbf{H}_{IC})_2 - \mathbf{H}_{IC})_1 \qquad (a)$$

The moment M about I_{IC} is:

$$M_{IC} = F(r_{In} + r_{Out}) = 120(0.5 + 1.0) = 120(1.5) = 180 \text{ N-m} \qquad (b)$$

The angular impulse $\mathbf{IA}_{IC}$ due to M_{IC} is:

$$\mathbf{IA}_{IC} = M_{IC} t = 180(5) = 900 \text{ N-m-s} \qquad (c)$$

The angular momentum about IC is:

$$H_{IC2} = I_{IC} \omega \qquad (d)$$

Let's use the parallel axis theorem to determine I_{IC}:

$$I_{IC} = I_G + mr^2 = mk_G^2 + mr^2 = m(k_G^2 + r^2) = 125[(0.45)^2 + (1.0)^2] = 150.3 \text{ kg-} m^2$$

Equating the angular impulse from Eq. (c) and the angular momentum from Eq. (d) yields:

$$900 = 150.3 \omega \qquad \text{hence} \qquad \omega = 5.988 \text{ rad/s} \qquad (e)$$

The velocity v_G of the disk is:
$$v_G = r_{Out} \omega = 1.0(5.988) = 5.988 \text{ m/s} \qquad (f)$$

EXAMPLE 9.4

A disk, initially at rest with a mass of 90 kg, is subjected to a moment given by $M = 4t^2$. The disk has a radius of gyration about G of $k_G = 0.4$ m. Determine the angular velocity of the disk, after the moment has been applied for 3.5 s.

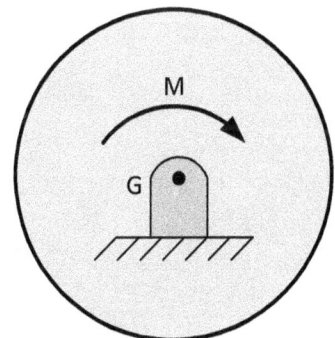

Fig. E9.4 A disk subjected to a moment $M = 4t^2$.

Solution:

We will employ the relations between angular impulse and angular momentum to determine the angular velocity of the disk, after the moment has been applied for 3.5 s. Because the disk is initially at rest, the value of the initial angular momentum is zero.

The angular impulse is:

$$\mathbf{IA} = \int_0^{3.5} M dt = \int_0^{3.5} 4t^2 dt = \frac{4}{3}\left[t^3\right]_0^{3.5} = \frac{4}{3}(3.5)^3 = 57.17 \text{ N - m - s} \qquad \text{(a)}$$

The final angular momentum is given by Eq. (9.12) as:

$$H_G = I_G \,\omega = mk_G^2 \,\omega = 90(0.4)^2 \,\omega = 14.4 \,\omega \qquad \text{(b)}$$

Equating the angular impulse and angular momentum yields:

$$14.4 \,\omega = 57.17 \qquad\qquad \omega = 3.970 \text{ rad/s} \qquad \text{(c)}$$

EXAMPLE 9.5

The pinion A and gear B are initially at rest. A constant moment M = 20 N-m is applied to the pinion, as shown in Fig. E9.5. The diameter of the pinion is 0.5 m and the diameter of the gear is 1.0 m. The radius of gyration of the pinion is $k_A = 0.12$ m and that of the gear is $k_B = 0.28$ m. The mass of the pinion is m_A = 25 kg m and that of the gear is m_B = 125 kg. Determine the angular velocity ω_B of gear B, after the moment has been applied for 4 s.

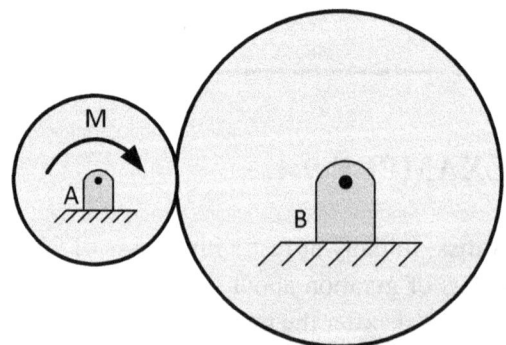

Fig. E9.5 The pinion A with an applied moment M is driving gear B.

Solution:

We will employ the relations between angular impulse and angular momentum to determine the angular velocity ω_B of gear B, after the moment has been applied for 4 s.
Let's prepare a free body diagram of the pinion A and gear B.

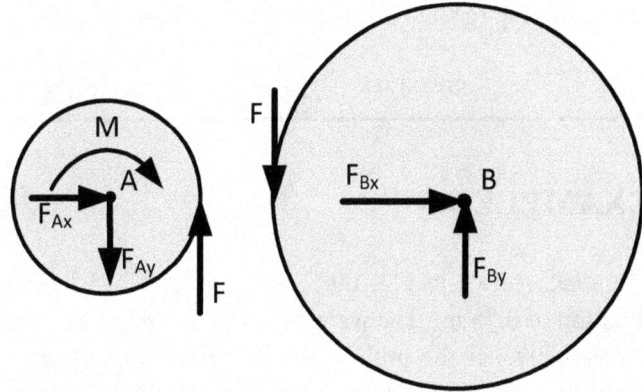

Fig. E 9.5a Free body diagrams of the gear and pinion. We have neglected the weighs of the pinion and gear in the FBDs, because they do not enter into the solution.

It is evident that the pinion A has an applied moment M and a tangential force F at the gear tooth, which is engaging gear B. Gear B has an equal and opposite force F at its mating tooth.

The angular impulse of the pinion A is given by Eqs. (9.4) and (9.18) as:

$$\mathbf{IA} = Mt - \int_0^4 0.25Fdt = 20(4) - \int_0^4 0.25Fdt \qquad (a)$$

The angular momentum of the pinion is given by Eq. (9.9) as:

$$H_A = I_A\,\omega_A = m_Ak_A^2\,\omega_A = 25(0.12)^2\,\omega_A = 0.36\,\omega_A \qquad (b)$$

Because of the diameters of the gears, we know that $\omega_A = 2\omega_B$ and write:

$$H_A = 0.72\,\omega_B \qquad (c)$$

The angular impulse applied to gear A is equal to its angular momentum. Accordingly, we equate Eq. (a) and Eq. (c) to obtain:

$$20(4) - \int_0^4 0.25Fdt = 0.72\,\omega_B \qquad (d)$$

The angular impulse applied to gear B is given by Eq. (9.18) as:

$$\mathbf{IA_B} = \int_0^4 M_B dt = \int_0^4 0.5Fdt \qquad (e)$$

The angular momentum of gear B is given by Eq. (9.9) as:

$$H_B = \mathbf{I_B}\,\omega_B = m_B\,k_B^2\,\omega_B = 125(0.28)^2\,\omega_B = 9.8 \text{ kg-m}^2/s \qquad (f)$$

Because the angular momentum H_B and the angular impulse IA_B are equal we obtain:

$$\int_0^4 0.5Fdt = 9.8 \qquad\qquad (g)$$

Substitute Eq. (g) into Eq. (d) and simplify to obtain:

$$80 - 4.9 = 0.72\,\omega_B \qquad\qquad \omega_B = 75.1/0.72 = 104.3 \text{ rad/s} \qquad (h)$$

EXAMPLE 9.6

The disk, shown in Fig. E9.6, has a mass of 15 kg and a radius of gyration of 0.75 m. The weights A and B have masses of 8 and 12 kg respectively. If the initial velocity of the weights is $v_A = 2.5$ m/s, determine the time required for the velocity of the two masses to increase until $v_B = 5.0$ m/s. Note mass A is moving upward and mass B is moving downward. Assume the cable employed to connect blocks A and B is massless.

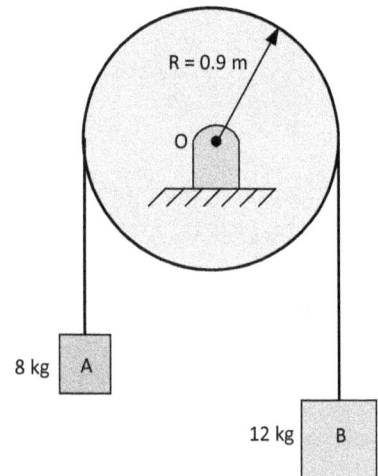

Fig. E9.6 A disk rotates about a fixed axis at point O.

Solution:

We will use angular momentum and angular impulse in our solution for this example. Note that the angular impulse added to the initial angular momentum gives the final angular momentum.

The initial angular momentum is:

For the disk $\qquad I_O \times \omega_1 = m\,k^2 v_1/r = 15(0.75)^2\,(2.5/0.9) = 23.44$ kg-m^2/s
For mass A $\qquad 8v_1 \times r = 8\,(2.5)(0.9) = 18$ kg-m^2/s
For mass B $\qquad 12v_1 \times r = 12\,(2.5)(0.9) = 27$ kg-m^2/s

The total initial angular momentum is: $\qquad\qquad 68.44$ kg-m^2/s

The initial angular impulse is:

For the system $\qquad [m_B - m_A]\,g\,r\,t = (12 - 8)(9.807)(0.9)\,t = 35.31\,t$ kg-m^2/s

The final angular momentum is:

For the disk $\qquad I_O\,\omega_2 = mk^2v_2/r = 15(0.75)^2\,(5.0/0.9) = 46.88$ kg-m^2/s

For mass A $\qquad 8v_2 \times r = 8 \, (5.0)(0.9) = 36 \text{ kg-m}^2/\text{s}$

For mass B $\qquad 12v_2 \times r = 12 \, (5.0)(0.9) = 54 \text{ kg-m}^2/\text{s}$

The total final angular momentum is: $\qquad 136.88 \text{ kg-m}^2/\text{s}$

Add the total initial angular momentum to the initial angular impulse and equate this result to the final angular momentum to obtain:

$$68.44 + 35.31 \, t = 136.88 \qquad \text{then} \qquad t = 68.44/35.31 = 1.938 \text{ s}$$

EXAMPLE 9.7

A weight B is connected with weightless cable to a mass $m_A = 75$ kg. The cable wraps around a disk with a radius of 0.8 m. The disk has a mass moment of inertia $I_G = 50$ kg-m^2. The block A slides on a horizontal surface, with a friction coefficient $\mu = 0.18$. The initial velocity of blocks A and B is $v_1 = 3$ m/s. After an interval of 5 s, the velocity of block B increases to $v_2 = 6$ m/s. Determine the mass of block B in order to achieve this increase in the velocity in 5 s.

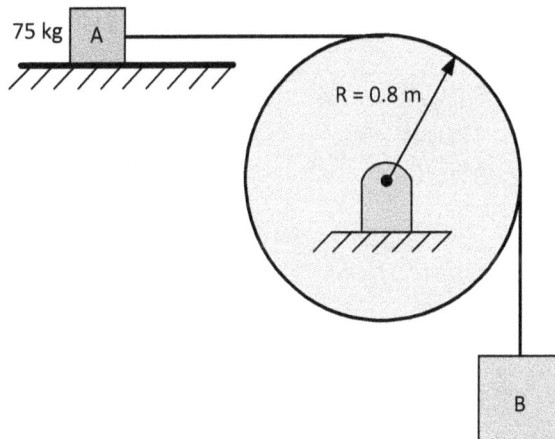

Fig. E9.7 A massless cable connecting blocks A and B is wrapped over a 90° arc of the disk.

Solution:

Let's begin by drawing free body diagrams of the three components, as shown below: We have neglected the pin forces and the weight in the FBD of the disk, because they do not enter into the solution.

By inspection, we establish the normal force N between Block A and the horizontal surface is:

$$N = W = m_A \, g = 75(9.807) = 735.5 \text{ N} \tag{a}$$

The friction force F_f opposing the motion of Block A is:

$$F_f = \mu \, N = 0.18(735.5) = 132.4 \text{ N} \tag{b}$$

The angular velocities of the disk at the initial and final time are:

$$\omega_1 = v_1/r = 3/0.8 = 3.75 \text{ rad/s}$$

$$\omega_2 = v_2/r = 6/0.8 = 7.50 \text{ rad/s} \tag{c}$$

The final angular momentum is equal to the initial angular momentum plus the angular impulse.

The initial angular momentum is:

Disk:	$I_G \, \omega_1 = 50(3.75) = 187.5 \text{ kg-m}^2/\text{s}$
Mass A	$m_A \, v_1 \, r = 75(3)(0.8) = 180 \text{ kg-m}^2/\text{s}$
Mass B	$m_B \, v_1 \, r = (3)(0.8) \, m_B = 2.4 \, m_B \text{ kg-m}^2/\text{s}$

$$\Sigma = 367.5 + 2.4 \, m_B \text{ kg-m}^2/\text{s} \tag{d}$$

The angular impulse is:

$$m_B \, g \, r \, t - F_f \, r \, t = [(9.807)(0.8) \, m_B - 132.4(0.8)] \, 5 = (39.23 \, m_B - 529.6) \tag{e}$$

The final angular momentum is:

Disk:	$I_G \, \omega_2 = 50(7.50) = 375.0 \text{ kg-m}^2/\text{s}$
Mass A	$m_A \, v_2 \, r = 75(6)(0.8) = 360 \text{ kg-m}^2/\text{s}$
Mass B	$m_B \, v_2 \, r = (6)(0.8) \, m_B = 4.80 \, m_B \text{ kg-m}^2/\text{s}$

$$\Sigma = 735.0 + 4.80 \, m_B \text{ kg-m}^2/\text{s} \tag{f}$$

Add the total initial angular momentum to the initial angular impulse and equate this result to calculate the final angular momentum:

$$367.5 + 2.4 \, m_B + 39.23 \, m_B - 529.6 = 735.0 + 4.80 \, m_B$$

$$36.83 \, m_B = 897.1 \qquad\qquad m_B = 24.36 \text{ kg} \tag{g}$$

9.4 IMPACT

9.4.1 Direct Central Impact

Impact is a topic where we must deviate from the concept of rigid bodies. Two bodies impacting involve local deformation at the point of contact. The local deformation can be either elastic or plastic. Small volumes, at the site of the impact on both bodies, are deformed and strain energy is briefly stored in the impacting bodies. During the impact, a contact force F acts for a very short period of time creating an impulse. This force causes the bodies to deform in a small region about the point of impact. The bodies then separate due to a recovery force R, and the strain energy is released. As the bodies separate, the negative impulse due to the recovery force R cancels out the positive impulse due to the impacting force F, the two bodies separate and the strain energy is released. The strain energy released influences the velocities of the bodies following the impact. In some cases, the velocities associated with the impact are sufficient to produces stresses larger than the yield strength of the material from which the impacting bodies are fabricated. In this situation, the impact produces plastic deformation, energy is lost and the change in the velocities following impact is more pronounced. Impact is a complex subject and the coverage here is limited to the elementary concepts.

Consider two spheres (particles) impacting, as shown in Fig. 9.5. We define the line of impact, as the common normal to the surfaces in contact during the impact process. Because the line of impact coincides with the center of masses of the two spheres, we identify this situation as direct central impact. We define the plane of contact, as the plane passing through the point of contact and perpendicular to the line of contact, as shown in Fig. 9.5.

Fig. 9.5 Two bodies impacting along a line of impact that passes through both of their centers of gravity.

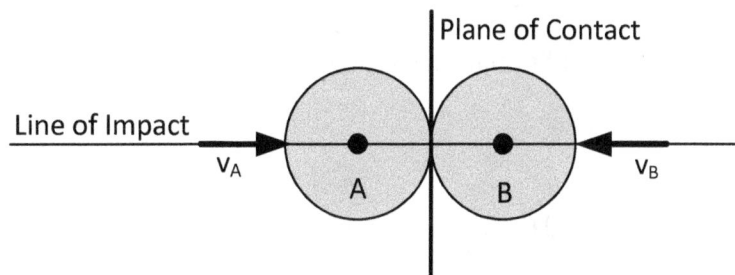

In dynamics we bodies to be rigid; however, when studying impact we make an exception and recognize that a small region of deformation occurs at the point of contact. We show an exaggerated deformed region on an impacted sphere in Fig. 9.5a. The line of impact is normal to this deformed surface.

Fig. 9.5a The localized region of deformation formed by impacting spheres and the line of impact that is normal to this surface.

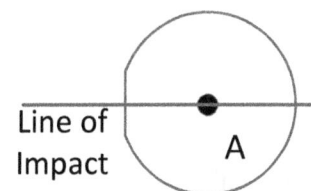

With direct central impact, for example, the two spheres separate after contact and move along the line of contact in the same direction, as shown in Fig. 9.6. Momentum is conserved for the system during impact, assuming that there is not a constraining surface that produces a reaction impulse to the system along the line of impact.

Fig. 9.6 After impact, the two bodies separate, with velocities that differ from their initial values.

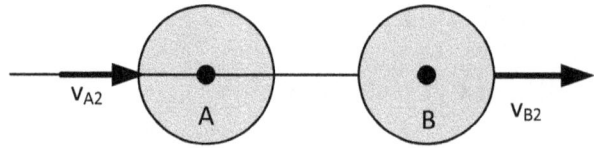

Let's consider conservation of momentum of the two spheres during the impact process. The linear momentum of the two spheres before and after impact is given by:

$$m_A \, v_{A1} + m_B \, v_{B1} = m_A \, v_{A2} + m_B \, v_{B2} \tag{9.19}$$

Let's examine the impact process more closely and note that the momentum and the impulse for sphere A during the deformation phase of the impact process are:

$$m_A \, v_{A1} - \int F \, dt = m_A \, v_M \tag{a}$$

where F is the force deforming sphere A and v_M is the velocity of sphere A when the deformation is a maximum.

The momentum and the impulse for sphere A during the recovery[1] phase of the impact process is:

$$m_A \, v_M - \int R \, dt = m_A \, v_{A2} \tag{b}$$

where R is the force deforming sphere A during the recovery phase.

The coefficient of restitution e is defined as:

$$e = \frac{\int R dt}{\int F dt} \tag{9.20}$$

From Eqs. (a) and (b) it is clear that:

$$e = \frac{v_M - v_{A2}}{v_{A1} - v_M} \tag{c}$$

It can be shown that a similar equation for the coefficient of restitution for sphere B is:

$$e = \frac{\int R dt}{\int F dt} = \frac{v_{B2} - v_M}{v_M - v_{B1}} \tag{d}$$

Cross multiply Eqs. (9.20) and (c), add the result to eliminate the unknown v_M and solve for e to obtain:

$$e = \frac{\int R dt}{\int F dt} = \frac{v_{B2} - v_{A2}}{v_{A1} - v_{B1}} \tag{9.21}$$

_____

[1] Often called the restitution phase of the impact event.

EXAMPLE 9.8

Two spheres of equal mass m impact, with the velocities that are shown in Fig. E9.8. The velocities of the spheres after impact are shown in Fig. E9.8a. Determine the coefficient of restitution.

Fig. E9.8 Two spheres prior to impact.

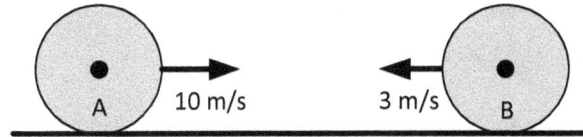

Fig. E9.8a Two spheres after impact.

Solution:

Define the velocities that are directed to the right as positive and those to the left as negative.

Substituting numerical values for the four velocities in Eq. (9.21) gives:

$$ e = \frac{v_{B2} - v_{A2}}{v_{A1} - v_{B1}} = \frac{6 - 2}{10 - (-3)} = 0.3077 $$

EXAMPLE 9.9

A ball with a mass of 1.5 kg is released from rest a distance of 1.8 m above a hard smooth surface, as shown in Fig. E9.9. The ball rebounds to a height of 1.4 m, determine the coefficient of restitution.

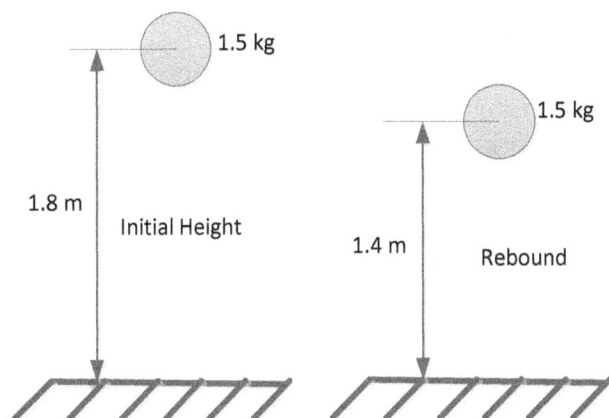

1.5 kg

1.5 kg

1.8 m

Initial Height

1.4 m

Rebound

Fig. E9.9 Rebound of a ball with a mass of 1.5 kg dropped on a hard surface.

Solution:

We will use the conservation of energy to determine the velocities v_{A1} and v_{A2}. Because the hard surface represents mass B that cannot move, we can set $v_{B1} = v_{B2} = 0$.

The initial kinetic energy $E_{k1} = 0$. The final kinetic energy $E_{k2} = \frac{1}{2} m v_{A1}^2$ (a)

The initial potential energy is $E_{P1} = Wy$ The final potential energy is $E_{P2} = 0$ (b)

Using the principle of the conservation of energy gives:

$$E_{k1} + E_{P1} = E_{k2} + E_{P2} = 0 + m(9.807)(1.8) = + \tfrac{1}{2} m v_{A1}^2 + 0 \qquad (c)$$

$$v_{A1} = 5.942 \text{ m/s} \qquad \text{downward at impact} \qquad (d)$$

After impact, we repeat the procedure using the principle of conservation of energy:

The initial kinetic energy $E_{k1} = \frac{1}{2} m v_{A2}^2$. The final kinetic energy $E_{k2} = 0$ (e)

The initial potential energy is $E_{P1} = 0$ The final potential energy is $E_{P2} = Wy$ (f)

Conservation of energy gives:

$$E_{k1} + E_{P1} = E_{k2} + E_{P2} = \tfrac{1}{2} m v_{A2}^2 + 0 = 0 + m(9.807)(1.4) \qquad (g)$$

$$v_{A2} = 5.240 \text{ m/s} \qquad \text{upward} \qquad (h)$$

$$e = \frac{v_{B2} - v_{A2}}{v_{A1} - v_{B1}} = \frac{0 - (-5.240)}{5.942 - 0} = 0.8819 \qquad (i)$$

In Eq. (i) we defined the velocity v_{A1} positive, because the sphere A moves downward and velocity v_{A2} negative as sphere A moves upward.

9.4.2 Oblique Impact

When two spheres collide at oblique angles, as shown in Fig. 9.7, they move away from the contact point with different velocities and at different angles. We define the line of impact as the common normal to the surfaces that are in contact during the impact process. We define the plane of contact as the plane passing through the point of contact and perpendicular to the line of contact.

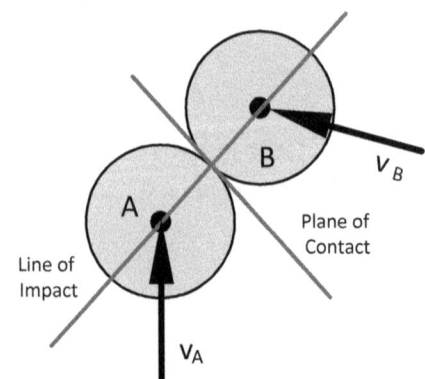

Fig. 9.7 Two spheres contacting with oblique impact.

With oblique impact of two particles, there are four known input parameters that include: the velocities v_{A1}, v_{B1} and the angles α_1 and β_1. There are also four unknown output parameters that include: the velocities v_{A2}, v_{B2} and the angles α_2 and β_2, as shown in Fig. 9.8.

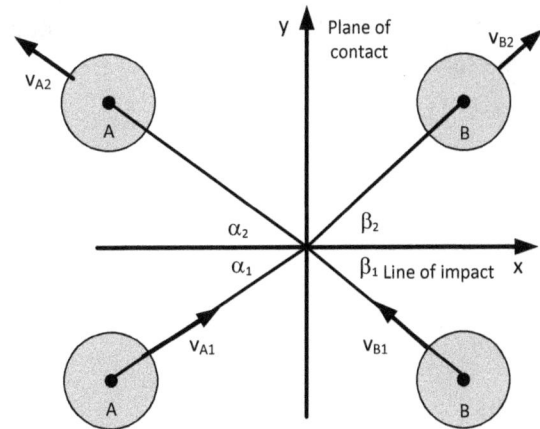

Fig. 9.8 Positions of spheres A and B before and after oblique impact.

A procedure to solve the oblique impact problem involves writing four different equations to solve for the four unknown output parameters. These equations employed are:

5. Momentum of the system is conserved along the line of impact (x axis). Hence:

$$m_A \, v_{A1} \cos \alpha_1 + m_B \, v_{B1} \cos \beta_1 = m_A \, v_{A2} \cos \alpha_2 + m_B \, v_{B2} \cos \beta_2 \qquad (9.22)$$

6. Momentum of sphere A is conserved along the plane of contact (y axis). Hence:

$$m_A \, v_{A1} \sin \alpha_1 = m_A \, v_{A2} \sin \alpha_2 \qquad \text{or} \qquad v_{A1} \sin \alpha_1 = v_{A2} \sin \alpha_2 \qquad (9.23)$$

7. Momentum of sphere B is conserved along the plane of contact (y axis). Hence:

$$m_B \, v_{B1} \sin \beta_1 = m_B \, v_{B2} \sin \beta_2 \qquad \text{or} \qquad v_{B1} \sin \beta_1 = v_{B2} \sin \beta_2 \qquad (9.24)$$

8. The coefficient of restitution is applied to the velocity components of spheres A and B, relative to the line of impact. Hence:

$$e = \frac{v_{B2} \cos \beta_2 - v_{A2} \cos \alpha_2}{v_{A1} \cos \alpha_1 - v_{B1} \cos \beta_1} \qquad (9.25)$$

The four equations (9.22) to (9.25) enable the solution of the four unknown parameters arising with oblique impact of two spheres. Let's consider an example to demonstrate this solution technique.

EXAMPLE 9.10

Sphere B impacts a rigid and smooth wall with an initial velocity $v_{B1} = 30$ m/s with an angle of $\beta_1 = 45^O$, as shown in Fig. E9.10. The coefficient of restitution e = 0.80. Determine the velocity v_{B2} and the angle β_2 of the sphere as it rebounds

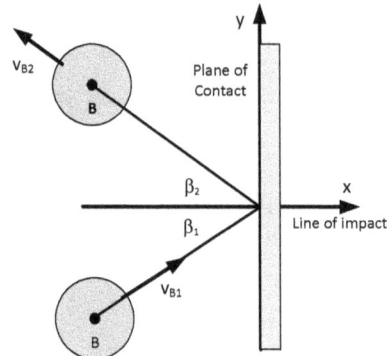

Fig. E 9.10 A sphere B strikes a rigid and smooth wall   and
rebounds with a new velocity and a new angle.

Solution:

Before attempting to employ the four equations developed to solve an oblique incidence impact exercise, recognize the mass of the wall approaches infinity, when compared to the mass of sphere B. The wall will not move; hence, $v_{A1} = v_{A2} = 0$. It is also evident that we cannot use either Eq. (9.22) or Eq. (9.23) in our solution, because the mass of the wall approaches infinity.

Because the momentum of sphere B is conserved along the plane of contact (y axis), we may use Eq. (9.24) and write:

$$v_{B1} \sin \beta_1 = 30 \sin 45^O = 21.21 \qquad v_{B2} \sin \beta_2 = v_{B1} \sin \beta_1 = 21.21 \qquad (a)$$

Equation (9.25) for the coefficient of restitution is applied to particle B relative to the line of impact to give:

$$e = \frac{v_{B2}\cos\beta_2 - 0}{0 - v_{B1}\cos\beta_1} = \frac{v_{B2}\cos\beta_2}{-30(0.7071)} = 0.80 \qquad (b)$$

$$v_{B2} \cos \beta_2 = -16.97 \qquad (c)$$

Equation (c) has a negative value, because the component of v_{B2} in the x direction is negative.

From Eqs. (a) and (c) we obtain:

$$\tan \beta_2 = 21.21/-16.97 = -1.2499 \qquad \beta_2 = -51.34^O \qquad (d)$$

From Eq. (a) we find:

$$v_{B2} = 21.21/ \sin \beta_2 = 21.21/-0.7809 = -27.16 \text{ m/s} \qquad (e)$$

Particle B is moving up and to the left in the positive y direction, but in the negative x direction; hence, the negative sign for v_{B2}.

9.5 ECCENTRIC IMPACT

In the previous discussion, we have considered centric and oblique impact of spheres that were treated as particles. In this section, we will develop methods for solving problems involving impact between two planar bodies, which usually involves eccentric impact. When the line of impact does not coincide with the line between the mass centers of the two impacting bodies, eccentric impact occurs. We illustrate eccentric impact in Fig. 9.9, where body A, translating with a velocity $\mathbf{u}_A$, strikes body B. Body B, supported by a clevis at point O, is rotating about that point with an angular velocity $\boldsymbol{\omega}_{B1}$. The velocity of body A along the line of contact v_A is a component of the velocity $\mathbf{u}_A$ and is dependent on the geometry of the contacting surfaces. Because body B is rotating with an angular velocity ω_{B1}, the linear velocity at the contact point Q, the instant before impact, is $v_{B1} = \omega_{B1} r$.

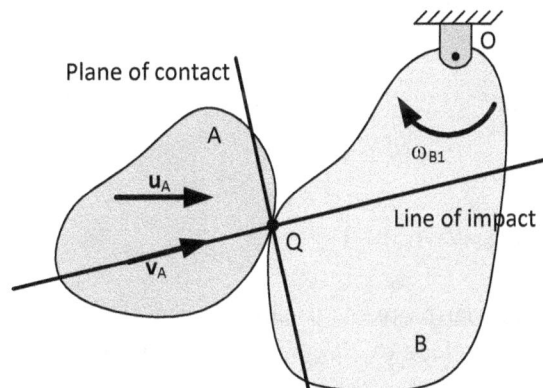

Fig. 9.9 Two planar bodies with eccentric impact occurring at point Q.

As before, we define the line of impact as the common normal to the surfaces in contact during the impact process. The plane of contact is the plane passing through the point of contact and perpendicular to the line of contact.

After impact, local deformation occurs at the contact point on both bodies, and equal and opposite impact forces F develop between the two bodies. These impact forces are directed along the line of impact, as shown in Fig. 9.10. At the conclusion of the deformation phase of the impact process, the two bodies are moving with the same component of velocity v_M along the line of impact.

During the second phase of the impact process, the local deformations in the small region about the contact point vanish. Equal and opposite impact forces R remain during this reconstitution process, as shown in Fig. 9.11. These forces act along the line of impact and additional forces develop at the pin located at point O. At the conclusion of the deformation and reconstitution phases of the impact process, the velocities of both bodies at the contact point Q are u_{A2} and v_{B2}, respectively.

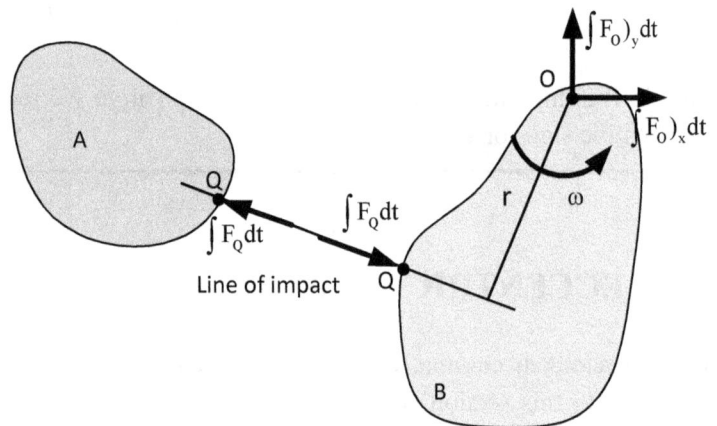

Fig. 9.10 Equal and opposite impact forces develop along the line of impact on both bodies, during the deformation phase of the impact process.

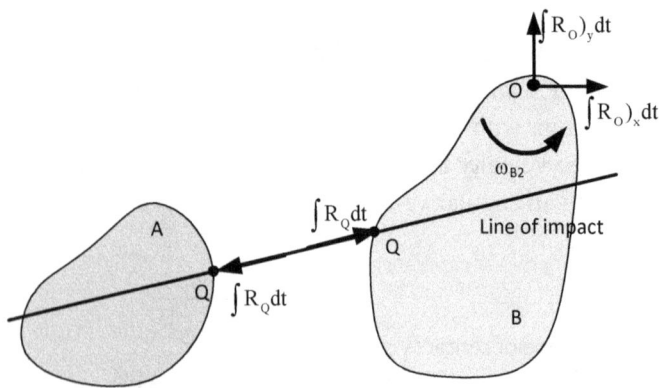

Fig. 9.11 Equal and opposite impact forces develop along the line of impact on both bodies, during the reconstitution phase of the impact process.

9.5.1 Analysis Method

Eccentric impact problems usually involve v_{A1} and v_{B1} as the two unknowns with v_{A2} and v_{B2}, as given velocities or velocities that can be calculated from data provided by using the methods introduced previously. Typical solutions make use of the principle of conservation of angular momentum of the two bodies to provide one of the two equations that are required. The second equation is obtained for the relationship for the coefficient of restitution, which is often cited in the problem statement.

The principle of angular momentum and impulse applied to body B enables us to write:

$$I_o \omega_{B1} + r \int F \, dt = I_o \omega \qquad (9.26)$$

where ω is the angular velocity of body B at the end of the deformation process, when the deformation is a maximum. I_O is the mass moment of inertia of body B relative to point O.

A similar relation for the angular momentum and impulse for the reconstitution phase yields:

$$I_o \omega + r \int R \, dt = I_o \omega_{B2} \qquad (9.27)$$

Recall the relation for the coefficient of restitution as:

$$e = \frac{\int R dt}{\int F dt} \tag{9.28}$$

Substituting results for $\int R dt$ and $\int F dt$ from Eqs. (9.26) and (9.27) into Eq.(9.28) yields:

$$e = \frac{r(\omega_{B2} - \omega)}{r(\omega - \omega_{B1})} = \frac{V_{B2} - V_M}{V_M - V_{B1}} \tag{9.29}$$

Following the same procedure for body A, we can derive:

$$e = \frac{V_M - V_{A2}}{V_{A1} - V_M} \tag{9.30}$$

Cross multiply the results from Eqs. (9.29) and (9.30) and add to obtain:

$$e = \frac{V_{B2} - V_{A2}}{V_{A1} - V_{B1}} \tag{9.31}$$

These relations, together with the methods previously introduced, enable the solution of eccentric impact problems. Let's begin by considering the problem of a bullet that becomes embedded, when it is fired into a wooden panel.

EXAMPLE 9.11

A bullet with a mass of 0.1 kg and a velocity of 600 m/s is fired into a wooden panel that has a mass of 12 kg, as shown in Fig. E9.11. The panel, initially at rest, is supported by a pin and clevis arrangement at point O. The bullet penetrates the wooden panel and becomes embedded near its bottom edge, at the location shown in Fig. E9.11. The time for the bullet to embed in the panel is 0.8×10^{-3} seconds. Determine the angular velocity of the panel immediately after the bullet becomes embedded and the reactive impulse forces acting at the pin O.

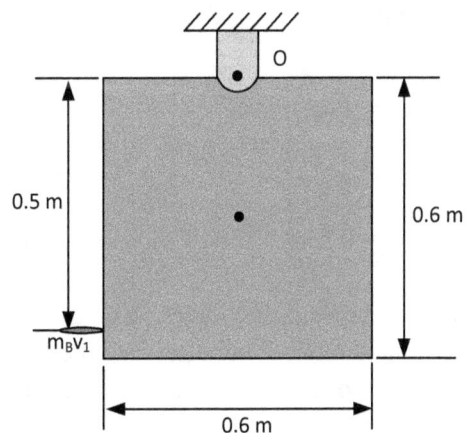

Fig. E9.11 A wooden panel suspended from a clevis is impacted by a bullet near its bottom.

Solution:

Because the bullet embeds in the panel, we define the bullet and the panel as a single body. During the impact process, the bullet and the panel develop equal and opposite internal impulses at the bullet's point of entry. The external impulses are due to the reactive forces at pin O, the weight of the panel and the bullet. The time of impact is very short, only 0.8 millisecond; hence, the movement of the panel is so small that the moments due to the weight of the panel and the bullet about point O have been neglected.

Let's prepare three drawings of the panel and the bullet in Fig. E9.12a, which show the initial system momentum, the external impulse and the final system momentum.

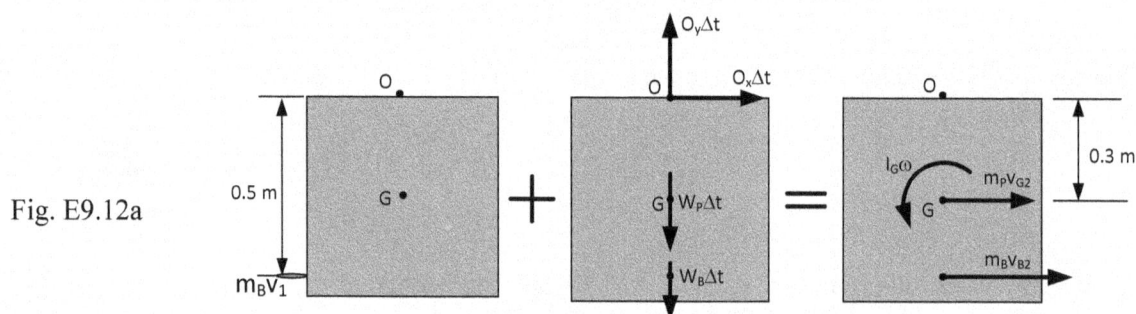

Fig. E9.12a

The principle of the conservation of angular momentum used with Eqs. (9.5) and (9.11) yields:

$$\mathbf{H}_{O1} = \mathbf{H}_{O2} \qquad (a)$$

$$0.5\, m_B\, v_B = 0.5 m_B\, v_{B2} + 0.3\, m_P\, v_{G2} + I_G\, \omega_2 \qquad (b)$$

$$I_G = (1/12) m_P\, [(a)^2 + (b)^2] = (1/12)(12)(2)(0.6)^2 = 0.72\ \text{kg-m}^2 \qquad (c)$$

Substituting numerical values into Eq. (b) yields:

$$0.5(0.1)(600) = 0.5(0.1) v_{B2} + 0.3(12) v_{G2} + 0.72\, \omega_2 \qquad (d)$$

Note that:
$$v_{B2} = 0.5\, \omega_2 \qquad\qquad v_{G2} = 0.3\, \omega_2$$

$$30 = (0.025 + 1.08 + 0.72)\, \omega_2 \qquad (e)$$

$$\omega_2 = 30/1.825 = 16.44\ \text{rad/s} \qquad (f)$$

$$v_{B2} = 0.5\, \omega_2 = 0.5(16.44) = 8.219\ \text{m/s} \qquad v_{G2} = 0.3\, \omega_2 = 0.3(16.44) = 4.932\ \text{m/s}$$

For reactive impulse forces, we use Eq. (9.1) to determine the linear momentum and Eq. (9.4) to determine the linear impulse. We then use the principle of conservation of linear momentum to write the expression for the x component of the force on the pin at impact.

$$m_B\, v_1 + O_x\, \Delta t = m_P\, v_{G2} \qquad (g)$$

$$0.1(600) + 0.8 \times 10^{-3} \, O_X = 12 \, (4.932)$$

$$O_x = -1,020 \, \text{N} \qquad\qquad\qquad\qquad \text{(h)}$$

In writing Eq. (g) we are determining the average value for O_x during impact. We have also assumed that the momentum of the bullet after it becomes embedded is negligible.

We follow the same procedure to determine the y component of the pin force as:

$$0 + O_y \, \Delta t - W_P \, \Delta t - W_B \, \Delta t = 0$$

$$O_y = W_P + W_B = (12 + 0.1)(9.807) = 118.7 \, \text{N} \qquad\qquad \text{(i)}$$

where W_P and W_B are the weights of the panel and bullet, respectively.

Next we will demonstrate an impact problem, where the projectile does not embed but rebounds.

EXAMPLE 9.12

A 3 kg sphere travelling horizontally with a velocity of 7.5 m/s impacts the end of a metallic rod, as shown in Fig. E9.12. The rod with a mass of 14 kg hangs from a clevis and is initially at rest. The coefficient of restitution between the sphere and the rod is 0.7. Determine the angular velocity of the rod and the velocity of the sphere immediately after impact.

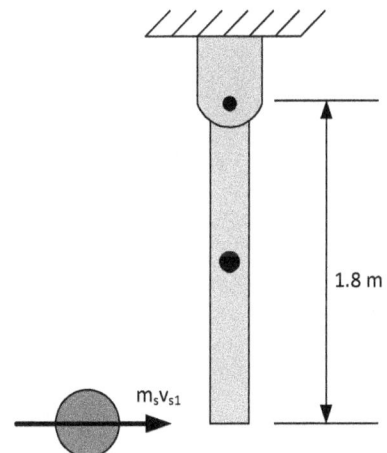

Fig. E9.12 A sphere with a velocity of 7.5 m/s impacts the end of a
metallic rod.

Solution:

We will divide the impact process into three phases, as shown in the sketch presented in Fig. E9.12. The first sketch shows the bar and sphere prior to impact. The second shows the reactions during impact, and the third sketch shows the bar and sphere after impact when the sphere rebounds.

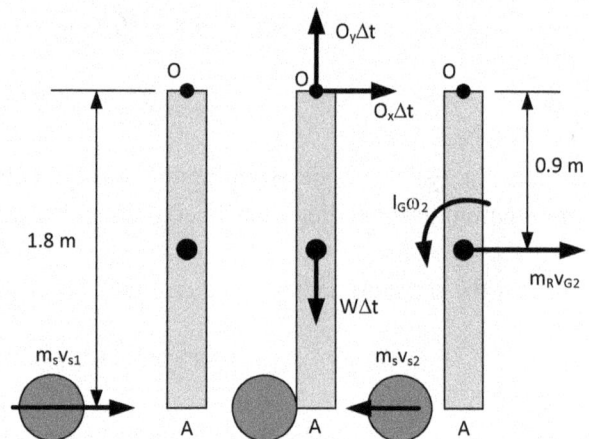

Fig. E9.12a. Three phases of the impact process.

In Fig. E9.12a we observe that there are no impact reaction forces at O, except during the impact phase. In reality there are reaction forces due to the weight of the bar; however, they have been neglected to emphasize the impact reaction forces.

Using the principle of the conservation of angular momentum, with Eqs. (9.5) and (9.11) yields:

$$1.8 \, m_S \, v_{S1} = 1.8 \, m_S \, v_{S2} + 0.9 m_R \, v_{G2} + I_G \, \omega_2 \qquad (a)$$

Note that
$$v_{G2} = 0.9 \, \omega_2 \qquad (b)$$

Recall
$$I_G = (1/12) m_R L^2 = (1/12)(14)(1.8)^2 = 3.78 \text{ kg-m}^2 \qquad (c)$$

Substitute Eq.(b) and (c) into Eq. (a) to give:

$$1.8(3)(7.5) = 1.8(3) v_{S2} + 0.9(14)(0.9)\omega_2 + 3.78 \, \omega_2$$

$$40.5 = 5.4 \, v_{S2} + 15.12 \, \omega_2 \qquad (d)$$

From Eq. (9.21) we write:

$$v_{A2} - v_{S2} = e \, (v_{S1} - v_{A1}) = 0.7(7.5 - 0) = 5.25 \text{ m/s} \qquad (e)$$

$$v_{A2} = 1.8 \, \omega_2 \qquad \text{and} \qquad v_{S2} = 1.8\omega_2 - 5.25 \qquad (f)$$

From Eqs. (d) and (f) we solve for ω_2 and obtain:

$$40.5 = 5.4(1.8 \, \omega_2 - 5.25) + 15.2 \, \omega_2 \qquad \omega_2 = 2.772 \text{ rad/s} \qquad (g)$$

Then from Eqs. (f) and (g) we write:

$$v_{S2} = 1.8(2.772) - 5.25 = -0.2604 \text{ m/s} \qquad (h)$$

The negative sign indicates that the sphere has rebounded from the rod and is moving to the left.

9.6   SUMMARY

We define the momentum **M** of the particle as:

$$\mathbf{M} = m\mathbf{v} \tag{9.1}$$

Newton's second law:

$$\Sigma\mathbf{F} = d(m\mathbf{v})/dt = d\mathbf{M}/dt \tag{9.2}$$

Integration yields:

$$\mathbf{I} = \int_{t_1}^{t_2} \sum \mathbf{F} dt = \int_{M_1}^{M_2} d\mathbf{M} \tag{9.3}$$

We defined the impulse **I** of the particle as:

$$\mathbf{I} = \int_{t_1}^{t_2} \sum \mathbf{F} dt \tag{9.4}$$

$$\mathbf{I} = \Delta\mathbf{M} = \mathbf{M_2} - \mathbf{M_1} \tag{9.4a}$$

The angular momentum $\mathbf{H}_O$ is:

$$\mathbf{H}_O = \rho \times \mathbf{M} = \boldsymbol{\rho} \times (m\mathbf{v}) \tag{9.5}$$

When the rigid body is translating, the momentum is given by:

$$\mathbf{M} = m\mathbf{v_G} \tag{9.6}$$

The angular momentum $\mathbf{H}_G$ relative to the center of gravity of a rigid body is:

$$\mathbf{H}_G = I_G\,\boldsymbol{\omega} = 0 \tag{9.7}$$

The linear momentum for fixed axis rotation is:

$$\mathbf{M} = m\,\mathbf{v}_G = m\,r_G\,\boldsymbol{\omega} \tag{9.8}$$

The angular momentum about the fixed point O is:

$$\mathbf{H}_O = I_O\,\boldsymbol{\omega} \tag{9.9}$$

$$I_O = I_G + m\,r_G^2 \tag{9.10}$$

The total angular momentum is:

$$\mathbf{H}_O = I_G\,\boldsymbol{\omega} + mr_G\,\mathbf{v}_G \tag{9.11}$$

$$H_G = I_G \, \omega \tag{9.12}$$

The angular momentum relative to a center of rotation O is:

$$\mathbf{H}_O = I_G \, \boldsymbol{\omega} + s \, m \, \mathbf{v}_G \tag{9.13}$$

$$\mathbf{H}_G = I_G \, \boldsymbol{\omega} \tag{9.14}$$

$$\Sigma M_G = I_G \, \alpha = I_G \, (d\omega/dt) \tag{9.15}$$

$$\Sigma \int_{t_1}^{t_2} M_G dt = I_G \int_{\omega_1}^{\omega_2} d\omega = I_G(\omega_2 - \omega_1) \tag{9.16}$$

Restricting the motion of the rigid body to rotation about a fixed axis located at some point O yields:

$$\Sigma \int_{t_1}^{t_2} M_O dt = I_O(\omega_2 - \omega_1) \tag{9.17}$$

The angular impulse is given by:

$$\mathbf{IA} = \Sigma \int_{t_1}^{t_2} M_G dt = I_G(\omega_2 - \omega_1) = \mathbf{H_G})_2 - \mathbf{H_G})_1 \tag{9.18}$$

The linear momentum of the two spheres before and after direct central impact is given by:

$$m_A \, v_{A1} + m_B \, v_{B1} = m_A \, v_{A2} + m_B \, v_{B2} \tag{9.19}$$

The coefficient of restitution is:

$$e = \frac{\int R dt}{\int F dt} \tag{9.20}$$

this can be written as:

$$e = \frac{\int R dt}{\int F dt} = \frac{v_{B2} - v_{A2}}{v_{A1} - v_{B1}} \tag{9.21}$$

With oblique impact

1. Momentum of the system is conserved along the line of impact (x axis).

$$m_A \, v_{A1} \cos \alpha_1 + m_B \, v_{B1} \cos \beta_1 = m_A \, v_{A2} \cos \alpha_2 + m_B \, v_{B2} \cos \beta_2 \tag{9.22}$$

2. Momentum of sphere A is conserved along the plane of contact (y axis).

$$m_A \, v_{A1} \sin \alpha_1 = m_A \, v_{A2} \sin \alpha_2 \qquad \text{or} \qquad v_{A1} \sin \alpha_1 = v_{A2} \sin \alpha_2 \qquad (9.23)$$

3. Momentum of sphere B is conserved along the plane of contact (y axis).

$$m_B \, v_{B1} \sin \beta_1 = m_B \, v_{B2} \sin \beta_2 \qquad \text{or} \qquad v_{B1} \sin \beta_1 = v_{B2} \sin \beta_2 \qquad (9.24)$$

4. The coefficient of restitution is applied to the velocity components of spheres A and B relative to the line of impact.

$$e = \frac{v_{B2} \cos \beta_2 - v_{A2} \cos \alpha_2}{v_{A1} \cos \alpha_1 - v_{B1} \cos \beta_1} \qquad (9.25)$$

With eccentric impact, the principle of angular momentum and impulse applied to an impacted body gives:

$$I_0 \omega_{B1} + r \int F dt = I_0 \omega \qquad (9.26)$$

A similar relation for the angular momentum and impulse for the reconstitution phase yields:

$$I_0 \omega + r \int R dt = I_0 \omega_{B2} \qquad (9.27)$$

Recall the relation for the coefficient of restitution as:

$$e = \frac{\int R dt}{\int F dt} \qquad (9.28)$$

$$e = \frac{r(\omega_{B2} - \omega)}{r(\omega - \omega_{B1})} = \frac{v_{B2} - v_M}{v_M - v_{B1}} \qquad (9.29)$$

For the impacting body A, we showed:

$$e = \frac{v_M - v_{A2}}{v_{A1} - v_M} \qquad (9.30)$$

From Eqs. (9.29) and (9.30) we showed:

$$e = \frac{v_{B2} - v_{A2}}{v_{A1} - v_{B1}} \qquad (9.31)$$

EQUATIONS FROM EACH CHAPTER

PART I KINEMATICS OF PARTICLES AND RIGID BODIES

Chapter 1 BASIC CONCEPTS IN MECHANICS

Newton's law in statics:

$$\Sigma \mathbf{F} = 0 \tag{1.2}$$

$$\Sigma \mathbf{F} = \mathbf{ma} \tag{1.3}$$

The force or weight due to gravity is:

$$F = m_b\, g = W \tag{1.7}$$

The force due to pressure is:

$$F = p\,A \tag{1.8}$$

Chapter 2 RECTILINEAR MOTION OF A PARTICLE

Kinematics deals with the motion of bodies without regard to the forces that are required to produce that motion.

Particles are small entities that may have mass but not physical dimensions. A particle is infinitesimally small when compared to distance it may travel.

Kinematics involves establishing the relations for the displacement, velocity and acceleration of a particle in motion using the time and geometry of the path of the particle. Forces required to produce this motion are not considered, although significant forces develop in order to produce the accelerations involved in certain problems.

When a particle moves in rectilinear motion, the position, velocity or acceleration are determined from:

$$v = \frac{dx}{dt} \tag{2.2}$$

$$a = \frac{dv}{dt} = \frac{d^2x}{dt^2} \tag{2.6}$$

$$a = v\frac{dv}{dx} \tag{2.9}$$

$$\int_{x_0}^{x} dx = \int_{0}^{t}\left[v_0 + \int_{0}^{t} f(t)dt \right]dt \tag{2.11}$$

If a particle is subjected to a constant acceleration a_0, then:

$$v = v_0 + a_0 t \tag{2.12}$$

$$x = x_0 + v_0 t + a_0 t^2/2 \tag{2.13}$$

$$v^2 = v_0^2 + 2a_0(x - x_0) \tag{2.14}$$

If a particle is traveling at a constant velocity,

$$a_0 = 0 \qquad \text{and} \qquad x = x_0 + v_c t \tag{2.15}$$

If the acceleration of a particle is a known function of position given by $a = f(x)$, then:

$$\frac{1}{2}\left(v^2 - v_0^2\right) = \int_{x_0}^{x} f(x)dx \tag{2.16}$$

If two particles P and Q are both moving along the x axis, their positions are given by x_P and x_Q. The distance between the two particles is $x_{Q/P}$ is:

$$x_{Q/P} = x_Q - x_P \tag{2.18a}$$

The velocity $v_{Q/P}$ is:

$$v_{Q/P} = v_Q - v_P \tag{2.19a}$$

The acceleration $a_{Q/P}$ is:

$$a_{Q/P} = a_Q - a_P \tag{2.20a}$$

Equations use in for numerical differentiation are:

$$\Delta y = y_2 - y_1 \tag{2.21}$$

$$\Delta t = t_2 - t_1 \tag{2.22}$$

then

$$\frac{\Delta x}{\Delta t} = \frac{x_2 - x_1}{t_2 - t_1} \tag{2.23}$$

The time corresponding to the derivative is:

$$t = (t_1 + t_2)/2 \tag{2.24}$$

Numerical integration involves summation of small rectangular areas A_n covering the total area A under a specified curve.

$$A = \sum_{n=1}^{N} A_n \tag{2.25}$$

Chapter 3 MOTION OF A PARTICLE IN THREE DIMENSIONS

The position of a particle moving along a curvilinear path is space is given by a position vector $\mathbf{r}$:

$$\mathbf{r} = x\,\mathbf{i} + y\,\mathbf{j} + z\,\mathbf{k} \tag{3.1}$$

The velocity vector $\mathbf{v}$ is determined by:

$$\mathbf{v} = dx/dt\,\mathbf{i} + dy/dt\,\mathbf{j} + dz/dt\,\mathbf{k} \tag{3.2}$$

where $\qquad v_x = dx/dt \qquad\qquad v_y = dy/dt \qquad\qquad v_z = dz/dt$

$$\mathbf{v} = v_x\,\mathbf{i} + v_y\,\mathbf{j} + v_z\,\mathbf{k} \tag{3.3}$$

The magnitude of the velocity vector is:

$$v = [v_x^2 + v_y^2 + v_z^2]^{1/2} \tag{3.4}$$

The direction of the velocity vector is expressed in terms of the unit tangent vector $\mathbf{u_v}$ as:

$$\mathbf{u_v} = \mathbf{v}/v \tag{3.5}$$

The acceleration of the particle is:

$$\mathbf{a} = d^2x/dt^2\,\mathbf{i} + d^2y/dt^2\,\mathbf{j} + d^2z/dt^2\,\mathbf{k} \tag{3.6}$$

where $\qquad a_x = d^2x/dt^2 \qquad\qquad a_y = d^2y/dt^2 \qquad\qquad a_z = d^2z/dt^2$

or $\qquad\qquad\qquad \mathbf{a} = a_x\,\mathbf{i} + a_y\,\mathbf{j} + a_z\,\mathbf{k} \tag{3.7}$

The magnitude of the acceleration vector is:

$$a = [a_x^2 + a_y^2 + a_z^2]^{1/2} \tag{3.8}$$

The direction of the acceleration vector is expressed in terms of a unit vector $\mathbf{u_a}$ as:

$$\mathbf{u_a} = \mathbf{a}/a \tag{3.9}$$

Planar motion is a special case of three dimensional curvilinear motion where:

$$\mathbf{r} = x\,\mathbf{i} + y\,\mathbf{j} \tag{3.10}$$

$$\mathbf{v} = dx/dt\,\mathbf{i} + dy/dt\,\mathbf{j} \tag{3.11}$$

$$\mathbf{v} = v_x\,\mathbf{i} + v_y\,\mathbf{j} \tag{3.12}$$

$$v = [v_x^2 + v_y^2]^{1/2} \tag{3.13}$$

$$\theta = \tan^{-1} v_y / v_x \tag{3.14}$$

$$\mathbf{a} = d^2x/dt^2 \, \mathbf{i} + d^2y/dt^2 \, \mathbf{j} \tag{3.15}$$

$$\mathbf{a} = a_x \, \mathbf{i} + a_y \, \mathbf{j} \tag{3.16}$$

$$a = [a_x^2 + a_y^2]^{1/2} \tag{3.17}$$

$$\theta = \tan^{-1} a_y / a_x \tag{3.18}$$

Projectile motion is a special case of planar curvilinear motion where:

The exit velocities of the projectile are:

$$v_x)_0 = v_{ex} \cos \theta \tag{3.19a}$$

and

$$v_y)_0 = v_{ex} \sin \theta \tag{3.19b}$$

After launch, the velocity of the projectile in the vertical (y) direction is:

$$v_y = v_y)_0 - g\,t \tag{3.20}$$

The relation for the height of the projectile, if the gun tube is initially located on the ground is:

$$y = v_y)_0\, t - gt^2/2 \tag{3.22}$$

The time required for the projectile to impact the ground plane is:

$$t_{Impact} = \frac{2v_y)_0}{g} \tag{3.23}$$

The range x is:

$$x = v_x)_0\, t_{Impact} \tag{3.24}$$

Equations for normal and tangential components of acceleration are:

The tangential velocity v_t is:

$$v_t = \rho\, \omega \tag{3.29}$$

The normal acceleration is:

$$a_n = v_t\, \omega = \rho\, \omega^2 = v_t^2/\rho \tag{3.30}$$

$$a_t = d^2s/dt^2 \tag{3.32}$$

$$a = [a_t^2 + a_n^2]^{1/2} \tag{3.33}$$

$$\theta = \tan^{-1} \frac{a_y}{a_x} \tag{3.34}$$

Equations for radial and transverse components of velocity and acceleration the position vector are:

$$\mathbf{r} = r\,\mathbf{u_r} \tag{3.35}$$

The velocity is:

$$\mathbf{v} = d\mathbf{r}/dt = r\,d\mathbf{u_r}/dt + dr/dt\,\mathbf{u_r} \tag{3.36}$$

The derivative of the unit vector $\mathbf{u_r}$ is:

$$\frac{d\mathbf{u_r}}{dt} = \lim_{t \to 0} \frac{\Delta\mathbf{u_r}}{\Delta t} = \lim_{t \to 0}\left(\frac{\Delta\theta}{\Delta t}\right)\mathbf{u_T} = \frac{d\theta}{dt}\mathbf{u_T} \tag{3.38}$$

The total velocity vector is:

$$\mathbf{v} = r\frac{d\theta}{dt}\mathbf{u_T} + \frac{dr}{dt}\mathbf{u_r} = v_T\mathbf{u_T} + v_r\mathbf{u_r} \tag{3.39}$$

where $v_T = r\,d\theta/dt = r\,\omega$ and $v_r = dr/dt$ and the magnitude of the velocity is given by:

$$v = [(v_r)^2 + (v_T)^2]^{1/2} \tag{3.40}$$

The velocity vector is tangent to the curvilinear path upon which the particle is moving.

The total acceleration vector is:

$$\mathbf{a} = d\mathbf{v}/dt = r\,d\theta/dt\,d\mathbf{u_T}/dt + d/dt(r\,d\theta/dt)\,\mathbf{u_T} + dr/dt\,d\mathbf{u_r}/dt + d^2r/dt^2\,\mathbf{u_r} \tag{3.41}$$

The derivative of the unit vector $\mathbf{u_T}$ is:

$$\frac{d\mathbf{u_T}}{dt} = \lim_{t \to 0} \frac{\Delta\mathbf{u_T}}{\Delta t} = -\lim_{t \to 0}\frac{\Delta\theta}{\Delta t}\mathbf{u_r} = -\frac{d\theta}{dt}\mathbf{u_r} \tag{3.42}$$

The total acceleration vector is:

$$\mathbf{a} = r\frac{d\theta}{dt}\left(-\frac{d\theta}{dt}\right)\mathbf{u_r} + \left(r\frac{d^2\theta}{dt^2} + \frac{d\theta}{dt}\frac{dr}{dt}\right)\mathbf{u_T} + \left(\frac{dr}{dt}\frac{d\theta}{dt}\right)\mathbf{u_T} + \frac{d^2r}{dt^2}\mathbf{u_r} \tag{3.43}$$

$$\mathbf{a} = a_r\,\mathbf{u_r} + a_T\,\mathbf{u_T} \tag{3.44}$$

$$a_r = [d^2r/dt^2 - r(d\theta/dt)^2] \qquad \text{and} \qquad a_T = [r\,d^2\theta/dt^2 + 2(d\theta/dt)(dr/dt)] \tag{3.45}$$

$$a = \{[d^2r/dt^2 - r(d\theta/dt)^2]^2 + [r\, d^2\theta/dt^2 + 2(d\theta/dt)(dr/dt)]^2\}^{1/2} \tag{3.46}$$

The angular acceleration is:

$$\alpha = \frac{d\omega}{dt} = \frac{d^2\theta}{dt^2} \tag{3.47}$$

Equations for particle motion along a circular path are:

$$\omega = \omega_0 + \alpha t \tag{3.49}$$

$$\theta = \omega_0 t + \tfrac{1}{2}\alpha t^2 \tag{3.50}$$

$$\omega^2 = \omega_0 + 2\alpha\theta \tag{3.51}$$

$$a_t = \frac{d}{dt}(r\omega) = r\frac{d\omega}{dt} \tag{3.52}$$

$$a_t = r\,\alpha \tag{3.53}$$

Chapter 4 KINEMATICS OF RIGID BODIES IN PLANAR MOTION

For a rigid bodies undergoing planar motion, a position vector $\mathbf{r_P}$ is defined that identifies a point in a body relative to a fixed coordinate system OXY. Then a moving coordinate system Qxy is established with a position vector $\mathbf{r_Q}$ that locates Q relative to a second set of coordinates OXY. Finally, the point P is located relative to Q with the position vector $\mathbf{r_{P/Q}}$ and expressed $\mathbf{r_P}$ as:

$$\mathbf{r_P} = \mathbf{r_Q} + \mathbf{r_{P/Q}} \tag{4.1}$$

Differentiating $\mathbf{r_P}$ with respect to time yields the velocity vector $\mathbf{v_P}$:

$$\mathbf{v_P} = d\mathbf{r_P}/dt = \mathbf{v_Q} + r_{P/Q}\,\omega\,\mathbf{u_\beta} \tag{4.2}$$

Differentiating $\mathbf{v_P}$ with respect to time yields the acceleration vector $\mathbf{a_P}$:

$$\mathbf{a_P} = d\mathbf{v_P}/dt = \mathbf{a_Q} - r_{P/Q}\,\omega^2\,\mathbf{u_r} + r_{P/Q}\,\alpha\,\mathbf{u_\beta} \tag{4.4}$$

where $\mathbf{u_r}$ and $\mathbf{u_\beta}$ are unit vectors in the r and β directions, respectively.

The angular velocity ω and the angular acceleration $\boldsymbol{\alpha}$ are given by:

$$\boldsymbol{\omega} = \mathbf{k} = \omega\,\mathbf{k} \quad\text{and}\quad \boldsymbol{\alpha} = d\omega/dt\,\mathbf{k} = \alpha\,\mathbf{k} \tag{4.5}$$

The velocity $\mathbf{v_P}$ is given by:

$$\mathbf{v_P} = \mathbf{v_Q} + \boldsymbol{\omega} \times \mathbf{r_{P/Q}} \tag{4.6}$$

The acceleration a_P may be expressed as:

$$\mathbf{a_P} = \mathbf{a_Q} + \boldsymbol{\omega} \times (\boldsymbol{\omega} \times \mathbf{r_{P/Q}}) + \boldsymbol{\alpha} \times \mathbf{r_{P/Q}} \tag{4.7}$$

Finally, you recognize that $\mathbf{v_P}$ and $\mathbf{a_P}$ can be rewritten as:

$$\mathbf{v_P} = \mathbf{v_Q} + \mathbf{v_{P/Q}} \tag{4.8}$$

$$\mathbf{a_P} = \mathbf{a_Q} + \mathbf{a_{P/Q}} \tag{4.9}$$

You treat fixed axis rotation, by locking the origin Q in place and rotating the body about a line through Q parallel to the Z axis. This constraint implies that $\mathbf{v_Q}$ and $\mathbf{a_Q}$ vanish. Hence:

$$\mathbf{v_P} = \boldsymbol{\omega} \times \mathbf{r_{P/Q}} \tag{4.14}$$

$$\mathbf{a_P} = \boldsymbol{\omega} \times (\boldsymbol{\omega} \times \mathbf{r_{P/Q}}) + \boldsymbol{\alpha} \times \mathbf{r_{P/Q}} \tag{4.15}$$

When a body is rotating about a fixed axis, the angular acceleration α is obtained by:

$$\alpha = d\omega/dt \tag{4.16}$$

$$\alpha = d^2\theta/dt^2 \tag{4.17}$$

The line of action of the angular acceleration coincides with the line of action of the angular velocity along the axis of rotation. The angular acceleration α is positive if ω is increasing and negative if ω is decreasing.

$$\alpha \, d\theta = \omega \, d\omega \tag{4.18}$$

If the angular acceleration is a constant α_k, integrate Eqs. (4.16), (4.17) and (4.18) to obtain:

$$\omega = \alpha_k \, t + \omega_0 \tag{4.19}$$

$$\theta = \tfrac{1}{2} \, \alpha_k t^2 + \omega_0 \, t + \theta_0 \tag{4.20}$$

$$\alpha_k \, (\theta - \theta_0) = \tfrac{1}{2} \, (\omega^2 - \omega_0^2)$$

or
$$\omega^2 = \omega_0^2 + 2 \, \alpha_k \, (\theta - \theta_0) \tag{4.21}$$

The relative velocity of two points on a rigid body is:

$$\mathbf{v_{P/Q}} = \mathbf{r_{P/Q}} \, \mathbf{d\theta/dt} = \mathbf{r_{P/Q}} \, \boldsymbol{\omega} \tag{4.22}$$

Because $\mathbf{v_{P/Q}}$ is due only to rotation, it may be expressed as a cross product:

$$\mathbf{v_{P/Q}} = \boldsymbol{\omega} \times \mathbf{r_{P/Q}} \tag{4.23}$$

or

$$v_P = v_Q + \omega \times r_{P/Q} \tag{4.24}$$

There is a significant advantage gained by identifying the instantaneous center IC of a body in motion. By placing the origin Q at the IC enables setting $v_Q = 0$ in Eq. (4.8) and writing:

$$v_P = v_{P/Q} = \omega \times r_{P/Q} \tag{4.25}$$

With point Q at an instantaneous center, place point P at any location on the body and compute the velocity at that point by using the cross product $\omega \times r_{P/Q}$. The angular velocity ω is usually known and the vector $r_{P/Q}$ is defined by the location of the point P.

The acceleration of point P can be rewritten as:

$$a_P = a_Q + (a_{P/Q})_n + (a_{P/Q})_t \tag{4.26}$$

where

$$(a_{P/Q})_n = r_{P/Q}\,\omega^2 \tag{4.27}$$

and

$$(a_{P/Q})_t = r_{P/Q}\,\alpha \tag{4.28}$$

The direction of the normal component of acceleration is from P to Q and the direction of the tangential component is perpendicular to $r_{P/Q}$. It is useful to express the acceleration of point P as:

$$a_P = a_Q - \omega^2\,r_{P/Q} + \alpha \times r_{P/Q} \tag{4.29}$$

PART II KINETICS OF PARTICLES AND RIGID BODIES

Chapter 5 KINETICS OF PARTICLE MOTION

With a particle, consider moments and rotation do not need to be considered, because the motion involves only translation, which is controlled by Newton's second law:

$$\Sigma F = m\frac{dv}{dt} = ma \tag{5.1}$$

The motion is referenced to an inertial coordinate system Oxyz, which is fixed in space.

Newton's second law and his gravitational law are:

$$F = m\,g = W \tag{5.2}$$

$$g = GM/R_e^2 \tag{5.3}$$

If the particle is Earth bound, g is the gravitational constant, W is the weight of the body and g is related to the universal gravitational constant G, the mass of the earth (M), and the radius of the Earth R_e.

In the International System of Units (SI), the mass is given in kilograms (kg) and the unit for force is a newton (N). In the U. S. Customary System, the mass is expressed in slugs and the unit for force is a pound (lb.)

There are three different coordinates systems that are used in dealing with particle motion, when the particle is subjected to applied forces.

For the rectangular coordinates, the equation of motion is expressed in a vector format as:

$$\Sigma F_x \, \mathbf{i} + \Sigma F_y \, \mathbf{j} + \Sigma F_z \, \mathbf{k} = m(a_x \, \mathbf{i} + a_y \, \mathbf{j} + a_z \, \mathbf{k}) \tag{5.4}$$

The scalar versions of the equations of motion in rectangular coordinates are:

$$\Sigma F_x = ma_x \qquad \Sigma F_y = ma_y \qquad \Sigma F_z = ma_z \tag{5.5}$$

If a particle is moving along a curvilinear path in the x-y plane, the normal and tangential components expressed in the equation of motion are:

$$\Sigma F_t \, \mathbf{u_t} + \Sigma F_n \, \mathbf{u_n} = m \, \mathbf{a_t} + m \, \mathbf{a_n} \tag{5.6}$$

where $\mathbf{u_t}$ and $\mathbf{u_n}$ are unit vectors in the tangential and normal directions, respectively and the tangential a_t and normal a_n components of acceleration are:

$$a_t = dv/dt \tag{5.7}$$

$$a_n = v^2/\rho \tag{5.8}$$

where ρ is the radius of curvature for the curvilinear path and the direction of the normal component of acceleration is toward the origin for ρ.

The scalar equations of motion in tangential and normal components of acceleration are:

$$\Sigma F_t = ma_t \qquad\qquad \Sigma F_n = ma_n \tag{5.9}$$

When a particle is moving along a curvilinear path in space, cylindrical coordinates are employed and the equation of motion is:

$$\Sigma F_r \, \mathbf{u_r} + \Sigma F_T \, \mathbf{u_T} + \Sigma F_z \, \mathbf{k} = m \, \mathbf{a_r} + m \, \mathbf{a_T} + m \, \mathbf{a_z} \tag{5.10}$$

where $\mathbf{u_r}$ and $\mathbf{u_T}$ and $\mathbf{k}$ are unit vectors in the radial, transverse and vertical directions respectively.

The components of acceleration in the radial and transverse directions are:

$$a_r = [d^2r/dt^2 - r(d\theta/dt)^2] \tag{5.11}$$

$$a_T = [r \, d^2\theta/dt^2 + 2(d\theta/dt)(dr/dt)] \tag{5.12}$$

The scalar equations of motion in cylindrical coordinates are:

$$\Sigma F_r = m\ a_r \qquad\qquad \Sigma F_T = m\ a_T \qquad\qquad \Sigma F_z = m\ a_z \qquad\qquad (5.13)$$

Chapter 6 KINETICS OF RIGID BODY MOTION

When the motion of a rigid body includes both translation and rotation, two equations of motion are employed in an analysis:

$$\Sigma \mathbf{F} = \mathbf{ma} \qquad\qquad \text{for translation} \qquad\qquad (6.1)$$

$$\Sigma \mathbf{M} = \mathbf{I\alpha} \qquad\qquad \text{for rotation} \qquad\qquad (6.2)$$

The mass moment of inertia is referenced to specified axes by:

$$I_z = \int_m r^2 dm = \int_m (x^2 + y^2)dm \qquad\qquad (6.3a)$$

$$I_x = \int_m r^2 dm = \int_m (y^2 + z^2)dm \qquad\qquad (6.3b)$$

$$I_y = \int_m r^2 dm = \int_m (x^2 + z^2)dm \qquad\qquad (6.3c)$$

The mass moment of inertia I in terms of a volume element dV is given by:

$$I = \rho \int_V r^2 dV \qquad\qquad (6.4)$$

where ρ is the mass density of the material used in fabricating the rigid body and $dm = \rho dV$.

The parallel axis theorem is:

$$I_z = \overline{I}_z + m(\overline{x}^2 + \overline{y}^2) \qquad\qquad (6.5)$$

where $\qquad \overline{I}_z = \int_m [x_G{}^2 + y_G{}^2]dm \qquad$ is the mass moment of inertia about the centroidal z axis.

$$I_x = \overline{I}_x + m(\overline{z}^2 + \overline{y}^2) \qquad\qquad (6.6)$$

$$I_y = \overline{I}_y + m(\overline{x}^2 + \overline{z}^2) \qquad\qquad (6.7)$$

$$I_z = \overline{I}_z + m\ d^2 \qquad\qquad (6.8)$$

where $\qquad\qquad d = \sqrt{(\overline{x}^2 + \overline{y}^2)} \qquad\qquad (6.9)$

In the equations listed above for the translation of rigid bodies, the distribution of the mass in the body was not considered. To consider the distribution of the mass of the body, rewrite the equation of motion:

$$\Sigma \mathbf{F} = m \, \mathbf{a}_G \qquad\qquad (6.12)$$

where the acceleration $\mathbf{a}_G$ refers to the center of mass of the rigid body.

The scalar version of the translation equations of motion is:

$$\Sigma F_x = m \, a_G)_x \qquad\qquad (6.13a)$$

$$\Sigma F_y = m \, a_G)_y \qquad\qquad (6.13b)$$

If the rigid body is translating without rotation:

$$\Sigma M_G = 0 \qquad\qquad (6.13c)$$

With curvilinear motion, but without rotation, the scalar representation of the equation of motion is:

$$\Sigma F_n = m \, a_G)_n \qquad\qquad (6.14a)$$

$$\Sigma F_t = m \, a_G)_t \qquad\qquad (6.14b)$$

$$\Sigma M_G = 0 \qquad\qquad (6.14c)$$

The equations are employed when analyzing rigid bodies in rectilinear motion.

For centroidal rotation of a plane rigid body subjected to a system of forces, the equation of motion is:

$$\Sigma \mathbf{F} = m\mathbf{a} = 0$$

The forces create a moment M_G that is given by:

$$\mathbf{M}_G = \Sigma \mathbf{r}_G \times \mathbf{F}$$

The moment vector $\mathbf{M}_G$ produces an angular acceleration of the body that is given by:

$$\mathbf{M}_G = I_G \, \boldsymbol{\alpha} \qquad\qquad (6.15)$$

Due to rotation, the tangential and normal components of the acceleration are:

$$a_t = r \, \alpha \qquad\qquad (6.16)$$

$$a_n = r \, \omega^2 \qquad\qquad (6.17)$$

With non-centroidal rotation, the tangential and normal components of acceleration are:

$$a_G)_t = \alpha\, r_G \tag{6.18}$$

$$a_G)_n = \omega^2\, r_G \tag{6.19}$$

When a plane body rotates about a non-centroidal axis, the center of mass moves along a circular path and the scalar equations of motion are given by:

$$\Sigma F_t = m\, a_G)_t = m\, \alpha\, r_G \tag{6.20}$$

$$\Sigma F_n = m\, a_G)_n = m\, \omega^2\, r_G \tag{6.21}$$

$$\Sigma M_G = I_G\, \alpha \tag{6.22}$$

The equation for the moments about a point A instead of point G is:

$$\Sigma M_A = (I_G + m r_G^2)\alpha = I_A\, \alpha \tag{6.23}$$

where:

$$I_A = I_G + m r_G^2 \tag{6.24}$$

In the general case of plane motion of rigid bodies, the body rotates and translates. The equations of motioned are referenced to an x-y axes and the moments are taken about the center of gravity.

$$\Sigma F_x = m\, a_G)_x \tag{6.25a}$$

$$\Sigma F_y = m\, a_G)_y \tag{6.25b}$$

$$\Sigma M_G = I_G\, \alpha \tag{6.25c}$$

The equations change somewhat with the motion of a point Q located some distance from the center of gravity G. The translation of the point Q and the translation of the center of gravity G are identical, hence:

$$\Sigma F_x = m\, a_G)_x \tag{6.26a}$$

$$\Sigma F_y = m\, a_G)_y \tag{6.26b}$$

The angular acceleration about point Q differs from the angular acceleration about the center of gravity G. The moment equation about an arbitrary point Q including the inertia forces $m\, a_G)_x$ and $m\, a_G)_y$ due to the linear accelerations:

$$\Sigma M_Q = \Sigma(\mathcal{M}_i)_Q = I_Q\, \alpha_Q \tag{6.26c}$$

$\Sigma(\mathcal{M}_i)_Q$ is given by:

$$\Sigma(\mathcal{M}_i)_Q = I_G\, \alpha_G + \overline{x}\, m\, a_G)_y + \overline{y}\, m\, a_G)_x \tag{6.27}$$

where ΣM_Q is due to the applied forces and moments; $\{\overline{x}\, [m\, a_G)_y] + \overline{y}\, [m\, a_G)_x]\}$ is the moment about Q due to the inertia forces. Note that $\overline{x}$ and $\overline{y}$ are the distances from point Q to the center of gravity G.

Chapter 7 WORK AND ENERGY: PARTICLE

The work E_W performed by a constant force F is:

$$E_W = F \int_{x_1}^{x_2} dx = F(x_1 - x_2) \tag{7.1}$$

If F is not a constant but a function of x, the work performed is:

$$E_W = \int_{x_1}^{x_2} f(x)dx \tag{7.2}$$

For a spring, with a spring rate k, the work to deform it is:

$$E_W = k \int_{x_1}^{x_2} x\,dx = (1/2)k\left[x_2^2 - x_1^2 \right] \tag{7.4}$$

Work is also performed by changing the elevation of a weight W:

$$E_W = mg(y_1 - y_2) = W\,(y_1 - y_2) \tag{7.5}$$

The work due to the change in elevation of the particle in a gravitational f field is equal to the potential energy E_P of the particle.

$$E_W = W\,(y_1 - y_2) = E_P)_1 - E_P)_2 \tag{7.6}$$

When the particle is moving along a curvilinear path, where the force F is a variable with r, the work E_W performed is given by the dot product:

$$E_W = \int_{r_1}^{r_2} F \cdot dr = \int_{s_1}^{s_2} F\cos\theta\,ds \tag{7.8}$$

The equation of motion for the motion of a particle in the tangential direction gives:

$$F_t = ma_t = m\,v\,dv/dx \tag{7.9}$$

The work performed is:

$$\int_{x_1}^{x_2} F_t dx = m\int_{v_1}^{v_2} v\,dv \tag{7.10}$$

The kinetic energy is:

$$E_k = \tfrac{1}{2}\,m(v_2^2 - v_1^2) \tag{7.12}$$

Energy is conserved, which enables us to write:

$$E_k)_1 + \Sigma E_W = E_k)_2 \tag{7.13}$$

Power P is defined as the rate of change of work E_W with respect to time.

$$P = dE_W/dt \qquad (7.14)$$

$$P = \mathbf{F} \cdot d\mathbf{r}/dt \qquad (7.15)$$

$$P = \mathbf{F} \cdot \mathbf{v} \qquad (7.16)$$

The power is a scalar quantity, because the dot product of two vectors yields a scalar result.

In the SI system of units, the power P is expressed in watts $\mathbf{W}$ where:

$$1 \mathbf{W} = 1 \text{ joule/s} = 1 \text{ J/s} = 1 \text{ N-m/s}$$

In the U. S. Customary system of units, the power P is expressed in horsepower HP, where:

$$1 \mathbf{HP} = 550 \text{ ft-lb/s} = 33,000 \text{ ft-lb/min.}$$

Efficiency is defined by:

$$\mathcal{E} = P_{Out}/P_{In} \qquad (7.17)$$

Efficiency may also be expressed in terms of energy E_{In} and energy E_{Out} of a system, if both are measured over the same time interval.

$$\mathcal{E} = E_{Out}/E_{In} \qquad (7.18)$$

Potential energy and work are equated to yield:

$$E_P = E_W = mg(y_1 - y_2) = W\, y \qquad (7.19)$$

The equation for potential energy stored in a spring is given by:

$$E_P = \tfrac{1}{2}\, kx^2 \qquad (7.20)$$

The principle of conservation of energy gives:

$$E_k)_1 + E_P)_1 = E_k)_2 + E_P)_2 \qquad (7.21)$$

Chapter 8 WORK AND ENERGY: RIGID BODIES

The kinetic energy dE_k of an incremental mass dm is:

$$dE_k = \tfrac{1}{2}\, dm\, v_Q^2 \tag{8.1}$$

Integrating yields:

$$E_k = 1/2 \int_m v_Q^2 dm \tag{8.2}$$

The velocity v_Q is given by:

$$v_Q = v_O + v_{Q/O} \tag{8.3}$$

$$v_Q = v_O)_x\, \mathbf{i} + v_O)_y\, \mathbf{j} + \omega\, \mathbf{k} \times (x\, \mathbf{i} + y\, \mathbf{j})$$

$$v_Q = [v_O)_x - \omega\, y]\, \mathbf{i} + [v_O)_y + \omega\, x]\, \mathbf{j} \tag{8.4}$$

$$v_Q^2 = v_O^2 - 2v_{O)x}\, \omega\, y + 2v_O)_y^2\, \omega\, x + \omega^2 r^2 \tag{8.5}$$

$$E_k = (1/2)v_O^2 - v_O)_x\, \omega\overline{y}m + v_O)_y\, \omega\overline{x}m + (1/2)I_O\omega^2 \tag{8.6}$$

If the origin of the Oxy coordinate system is taken at the mass center G of the rigid body, then:

$$E_k = (1/2)v_G^2 + (1/2)I_G\omega^2 \tag{8.7}$$

With fixed axis rotation about point O:

$$v_G = r_G\, \omega \tag{8.10}$$

$$E_k = (1/2)(mr_G^2 + I_G)\omega^2$$

$$E_k = (1/2)I_O\omega^2 \tag{8.11}$$

The kinetic energy in terms of the body's angular velocity about its instantaneous center is given by:

$$E_k = (1/2)I_{IC}\omega^2 \tag{8.12}$$

Work energy due to a variable force acting on a rigid body is given by:

$$E_W = \int F \bullet dr = \int_s F\cos\theta ds \tag{8.13}$$

Work energy due to a constant force F is given by:

$$E_W = (F \cos \theta)s \tag{8.14}$$

Work energy due the change in the elevation y of the weight W of a rigid body is given by:

$$E_W = W\, y = E_P \tag{8.15}$$

Work energy due the change in the length x of a spring in either tension or compression is given by:

$$E_W = \tfrac{1}{2}\, k(x_2{}^2 - x_1{}^2) \qquad (8.16a)$$

Work energy E_{WT} due the change in the angle of rotation of a torsion spring is given by:

$$E_{WT} = \tfrac{1}{2}\, k_T\, (\theta_2{}^2 - \theta_1{}^2) \qquad (8.16b)$$

Work energy is also developed due to moments rotating through an angle θ that is determined from:

$$dE_W = M\, d\theta$$

Integrating yields:

$$E_W = \int_{\theta_1}^{\theta_2} M\, d\theta \qquad (8.17)$$

If the moment M is a constant:

$$E_W = M(\theta_2 - \theta_1) \qquad (8.18)$$

Potential energy is computed relative to a datum plane. When the position of the weight y is above the datum plane the potential energy is positive and given by:

$$E_P = W\, y_+ \qquad (8.19)$$

When the position of the weight is below the datum plane the potential energy is negative and given by:

$$E_P = -\, W\, y_- \qquad (8.20)$$

Potential energy E_{PS} can be stored by the elastic deformation of a spring and is determined from:

$$E_{PS} = \tfrac{1}{2}\, k(x_2{}^2 - x_1{}^2) \qquad (8.21)$$

The total potential energy is due to both the elastic and the gravitational potential energy and is given by:

$$\Sigma E_P = E_P + E_{PS} \qquad (8.22)$$

Taking all the forms of energy into account in the initial and final states of motion yields:

$$E_{k1} + \Sigma E_W + E_{P1} = E_{k2} + E_{P2} \qquad (8.23)$$

Chapter 9 MOMENTUM AND IMPULSE: PARTICLES AND RIGID BODIES

The momentum **M** of the particle is defined as:

$$\mathbf{M} = m\mathbf{v} \tag{9.1}$$

Newton's second law:

$$\Sigma\mathbf{F} = d(m\mathbf{v})/dt = d\mathbf{M}/dt \tag{9.2}$$

Integration yields:

$$\mathbf{I} = \int_{t_1}^{t_2} \sum Fdt = \int_{M_1}^{M_2} d\mathbf{M} \tag{9.3}$$

The impulse **I** of a particle is defined as:

$$\mathbf{I} = \int_{t_1}^{t_2} \sum Fdt \tag{9.4}$$

$$\mathbf{I} = \Delta\mathbf{M} = \mathbf{M_2} - \mathbf{M_1} \tag{9.4a}$$

The angular momentum $\mathbf{H_O}$ is defined as:

$$\mathbf{H_O} = \rho \times \mathbf{M} = \rho \times (m\mathbf{v}) \tag{9.5}$$

When the motion of a rigid body is translation, the momentum is given by:

$$\mathbf{M} = m\mathbf{v_G} \tag{9.6}$$

The angular momentum $\mathbf{H_G}$ relative to the center of gravity of a rigid body is given by:

$$\mathbf{H_G} = I_G\,\omega = 0 \tag{9.7}$$

The linear momentum for fixed axis rotation is given by:

$$\mathbf{M} = m\,\mathbf{v_G} = m\,r_G\,\omega \tag{9.8}$$

The angular momentum about the fixed point O is given by:

$$\mathbf{H_O} = I_O\,\omega \tag{9.9}$$

$$I_O = I_G + m\,r_G^2 \tag{9.10}$$

The total angular momentum is given by:

$$H_O = I_G \, \omega + m \, r_G \, \mathbf{v}_G \qquad\qquad (9.11)$$

$$H_G = I_G \, \omega \qquad\qquad (9.12)$$

The angular momentum relative to a center of rotation O is given by:

$$\mathbf{H}_O = I_G \, \omega + s \, m \, \mathbf{v}_G \qquad\qquad (9.13)$$

$$\mathbf{H}_G = I_G \, \omega \qquad\qquad (9.14)$$

$$\Sigma M_G = I_G \, \alpha = I_G \, (d\omega/dt) \qquad\qquad (9.15)$$

$$\Sigma \int_{t_1}^{t_2} M_G dt = I_G \int_{\omega_1}^{\omega_2} d\omega = I_G (\omega_2 - \omega_1) \qquad\qquad (9.16)$$

Restricting the motion of the rigid body to rotation about a fixed axis located at some point O yields:

$$\Sigma \int_{t_1}^{t_2} M_O dt = I_O (\omega_2 - \omega_1) \qquad\qquad (9.17)$$

The angular impulse is given by:

$$I = \Sigma \int_{t_1}^{t_2} M_G dt = I_G (\omega_2 - \omega_1) = \mathbf{H_G})_2 - \mathbf{H_G})_1 \qquad\qquad (9.18)$$

The linear momentum of the two spheres before and after direct central impact is given by:

$$m_A \, v_{A1} + m_B \, v_{B1} = m_A \, v_{A2} + m_B \, v_{B2} \qquad\qquad (9.19)$$

The coefficient of restitution is given by:

$$e = \frac{\int R dt}{\int F dt} \qquad\qquad (9.20)$$

This relation can be written as:

$$e = \frac{\int R dt}{\int F dt} = \frac{v_{B2} - v_{A2}}{v_{A1} - v_{B1}} \qquad\qquad (9.21)$$

With oblique impact, these four statements provide the basis for solution of impact problems.

1. Momentum of the system is conserved along the line of impact (x axis).

$$m_A \, v_{A1} \cos \alpha_1 + m_B \, v_{B1} \cos \beta_1 = m_A \, v_{A2} \cos \alpha_2 + m_B \, v_{B2} \cos \beta_2 \qquad (9.22)$$

2. Momentum of sphere A is conserved along the plane of impact (y axis).

$$m_A \, v_{A1} \sin \alpha_1 = m_A \, v_{A2} \sin \alpha_2 \qquad \text{or} \qquad v_{A1} \sin \alpha_1 = v_{A2} \sin \alpha_2 \qquad (9.23)$$

3. Momentum of sphere B is conserved along the plane of impact (y axis).

$$m_B \, v_{B1} \sin \beta_1 = m_B \, v_{B2} \sin \beta_2 \qquad \text{or} \qquad v_{B1} \sin \beta_1 = v_{B2} \sin \beta_2 \qquad (9.24)$$

4. The coefficient of restitution is applied to the velocity components of spheres A and B relative to the line of impact.

$$e = \frac{v_{B2} \cos \beta_2 - v_{A2} \cos \alpha_2}{v_{A1} \cos \alpha_1 - v_{B1} \cos \beta_1} \qquad (9.25)$$

With eccentric impact the principle of angular momentum and impulse applied to an impacted body gives:

$$I_o \omega_{B1} + r \int F dt = I_o \omega \qquad (9.26)$$

A similar relation for the angular momentum and impulse for the reconstitution phase yields:

$$I_o \omega + r \int R dt = I_o \omega_{B2} \qquad (9.27)$$

The coefficient of restitution is given by:

$$e = \frac{\int R dt}{\int F dt} \qquad (9.28)$$

$$e = \frac{r(\omega_{B2} - \omega)}{r(\omega - \omega_{B1})} = \frac{v_{B2} - v_M}{v_M - v_{B1}} \qquad (9.29)$$

For the impacting body A, the coefficient of restitution is given by:

$$e = \frac{v_M - v_{A2}}{v_{A1} - v_M} \qquad (9.30)$$

INDEX